GUIDELINES
for the
CONTINUING MATHEMATICAL EDUCATION
of
TEACHERS
1988

The Mathematical Association of America
Committee on the Mathematical Education of Teachers
COMET

MAA COMMITTEE ON THE MATHEMATICAL EDUCATION OF TEACHERS

Henry L. Alder, Department of Mathematics, University of California, Davis, CA 95616.

Mary Kay Corbitt, 1724 Pitty Pat Court, Lilburn, GA 30247.

F. Joe Crosswhite, Department of Mathematics, Box 5717, Northern Arizona University, Flagstaff, AZ 86011 (from Jan. 1987).

John A. Dossey, Department of Mathematics, Illinois State University, Normal, IL 61761 (until Oct. 1985).

Marjorie Enneking, Department of Mathematics, Portland State University, Portland, OR 97207.

Richard K. Guy, Department of Mathematics, and Statistics, The University of Calgary, Calgary, Alberta, Canada, T2N 1N4. (editor until Jan. 1987).

Katherine P. Layton, Beverly Hills High School, Beverly Hills, CA 90212.

Calvin T. Long, Department of Pure and Applied Mathematics, Washington State University, Pullman, WA 99164 (co-chair, editor from May 1987).

Bruce E. Meserve, 521 South Paseo Del Cobre, Green Valley, AZ 85614 (co-chair).

Mary Ann Norton, Director of Mathematics, Spring Independent School District, 16717 Ella Blvd., Houston, TX 77090.

Alan Osborne, Department of Educational Theory and Practice, 202 Arps Hall, Ohio State University, Columbus, OH 43210.

ACKNOWLEDGEMENTS

Many others have provided useful input: they include David A. Bates, Rick Billstein, Gary G. Bitter, Christine Browning, Donald W. Bushaw, Gulbank D. Chakerian, Stanley R. Clemens, Sue Dolezal, Jean Gamlen, David K. Guy, John G. Harvey, Alan Hoffer, Zaven A. Karian, Donald L. Kreider, Patty Mitchell, Jean Pedersen, Gerald J. Porter, Anthony Ralston, John O. Riedle Jr., Fred S. Roberts, Jack M. Robertson, G. Thomas Sallee, Oscar Schaaf, Richard L. Scheaffer, Thomas L. Schroeder, Martha J. Siegel, Gerald D. Smetzer, Lynn Arthur Steen, Sherman K. Stein, Alan O. Tucker, Ann E. Watkins, Fred Weaver, Dorothy Williams, Ronald H. Wenger, and Hassler Whitney. Additional thanks are given to Louise and Richard Guy for providing the preliminary typescript and to the Department of Pure and Applied Mathematics at Washington State University for the final typescript.

iii

PREFACE

These GUIDELINES are a sequel to the RECOMMENDATIONS ON THE MATHEMATICAL PREPARATION OF TEACHERS (the 1983 RECOMMENDATIONS prepared by the CUPM Panel on Teacher Training and published in 1983 by the Mathematical Association of America as MAA Notes Number 2). Both the RECOMMENDATIONS and the GUIDELINES necessarily represent stages in the ongoing, and rapidly evolving, status of recommended programs for teachers. Periodic revisions of all such recommendations are essential in order to capitalize on changes in the pedagogical use of calculators, computers, and other technology, in the desirable emphasis in developing specific mathematical topics, in the appropriate selection of topics, and so on.

The 1983 RECOMMENDATIONS concern the mathematical preparation of prospective teachers. The present GUIDELINES are concerned with the continuing mathematical education of teachers as professionals who wish to upgrade or update their previous preparation. College and university departments of Mathematics and of Mathematics Education can participate in of the continuing education of teachers through

short courses designed for teachers,

inservice programs, both on and off campus,

advanced degree programs for teachers,

lectures at professional meetings for teachers,

and in other ways. These GUIDELINES are intended to encourage such activities and to enhance their effectiveness.

These GUIDELINES and the 1983 RECOMMENDATIONS represent efforts by the Mathematical Association of America to address the critical shortage of adequately qualified teachers of mathematics. In each case the sample courses and programs are intended to suggest promising approaches that may be modified to fit local situations. Types of programs, ways of meeting the needs of teachers, and general mathematical content are suggested. Details are expected to change in different situations and as additional resources and experiences are acquired.

CONTENTS

INTRODUCTION

The Mathematical Association of America (MAA) recognizes that the changing needs of both elementary and secondary school students require that inservice and preservice teachers have opportunities to meet contemporary standards of mathematical preparation. The Committee on the Undergraduate Program in Mathematics (CUPM) of the MAA has been concerned with the mathematical preparation of elementary and secondary school teachers throughout its thirty years of existence. In 1983, the CUPM Panel on Teacher Training published RECOMMENDATIONS ON THE MATHEMATICAL PREPARATION OF TEACHERS (MAA Notes Number 2). In what follows, these MAA Notes will be referred to as the 1983 RECOMMENDATIONS.

The MAA Committee on the Mathematical Education of Teachers (COMET) has replaced the CUPM Panel on Teacher Training. These GUIDELINES are the first report of COMET. This effort to improve the mathematical preparation of teachers is part of a broad movement to improve the teaching of mathematics. The MAA Notes No. 6, A LEAN AND LIVELY CALCULUS, Ronald C. Douglas (ed.), 1985, and MAA Notes No. 8, CALCULUS FOR A NEW CENTURY, Lynn Arthur Steen (ed.), 1988, indicate ongoing efforts to restructure the teaching of calculus. The Mathematical Sciences Education Board (MSEB) of the National Research Council is composed of organizations in mathematics education. Initially the MSEB is focusing on curriculum, instruction, and evaluation. The National Council of Teachers of Mathematics (NCTM) task force on the Post-Baccalaureate Education of Mathematics Teachers is preparing GUIDELINES FOR THE POST-BACCALAUREATE EDUCATION OF MATHEMATICS TEACHERS that specify competencies needed by teachers. The Commission on Standards for School Mathematics appointed by the NCTM is preparing CURRICULUM AND EVALUATION STANDARDS FOR SCHOOL MATHEMATICS that are expected to be released in April 1989.

INTENDED AUDIENCE

COMET addresses these GUIDELINES to:

- School administrators - to help plan inservice programs for their teachers and supervisors.

- University and college administrators - to design and staff courses for teachers.

- School teachers and supervisors - to help plan and to participate in inservice programs.

- University and college faculty members - to help design and conduct courses and inservice programs.

These GUIDELINES are also addressed to:

• National and state governmental agencies,

• Professional societies,

• Regional and local educational organizations,

and to everyone concerned with enhancing the effectiveness of mathematics teaching. These GUIDELINES provide a basis for the essential cooperation among these groups, as well as between departments of mathematics and departments of mathematics education in colleges and universities.

We start with recommendations for

inservice programs for all teachers.

The main part of these GUIDELINES consists of recommendations for

graduate programs for teachers.

who are certified at one of the levels specified in the 1983 RECOMMENDATIONS:

Level I   Teachers of elementary school mathematics,

Level II   Elementary school mathematics specialists, coordinators of elementary school mathematics programs, and teachers of middle school and junior high school mathematics, and

Level III Teachers of high school mathematics

Finally we consider recommendations for the

graduate education of mathematics supervisors.

Many institutions are committed to shifting initial certification programs from four to five year programs as recommended by the Holmes Group and in the report of the Carnegie Forum on Education and the Economy. Too much is known about teaching and learning to be covered thoroughly in a four-year program. The programs and courses described in these GUIDELINES are recommended for all previously certified teachers and are designed with the *inservice* teacher in mind to provide a fresh look at mathematics and methodology.

Course titles and names of degrees vary from institution to institution. No matter what the name of the degree, the program is designed specifically to address the needs of teachers. The academic standards of such programs, in terms of the challenge they present to the participant, are intended to be comparable with those of other master's degree programs. A minimum of thirty semester hours of graduate credit is assumed for each master's degree program.

ADVANTAGES OF INSERVICE AND MASTER'S DEGREE PROGRAMS

• **Schools**

  Make the learning of mathematics a more exciting experience for students.

  Provide opportunities for teachers to continue their professional growth.

  Obtain outside specialists for consultation and for assistance with enrichment activities.

• **University and college departments of mathematics and of mathematics education**

  Enhance their community relations through outreach programs.

  Identify and attract potential undergraduate and graduate students.

• **School teachers**

  Enhance their professional skills.

  Increase their opportunities for interaction with other professionals.

• **University and college faculty members**

  Obtain opportunities to influence school curricula.

  Contribute to the improvement of the mathematical preparation of future undergraduates.

Mathematics programs for teachers may be conducted by schools, school districts, colleges, universities, professional societies, business groups, or governmental agencies. These GUIDELINES are meant to encourage a wide variety of programs that are useful to teachers. Mathematics, pedagogical techniques, and the educational environment are all changing. Thus, these GUIDELINES indicate directions for continued program development rather than recommend details for particular present or future programs.

IMPORTANT BASIC PRINCIPLES

All courses must be exemplary models of effective teaching.

Use the best teaching strategies throughout your course, else your message falls on deaf ears. Teachers emulate their own favorite teachers.

If you don't enjoy teaching it, they won't enjoy learning it.

Convey that mathematics is fun, exciting, and full of surprises; that it's applicable in many different, often unexpected, ways.

Involvement of parents in their children's education is an extremely important factor affecting student achievement.

Successful strategies to involve parents need to be incorporated in courses for teachers.

The distinction between Mathematics courses and Mathematics Education courses is not always clear cut.

Either may be taught by Departments of Mathematics or of Curriculum and Instruction.

There must be constant communication among instructors of all courses.

Communication is especially important when separate Departments or Faculties are concerned.

The use of calculators and computers must be integrated into Mathematics and Mathematics Education courses wherever possible.

Instruction in, and use of, these skills is best included in content and methods courses. We list separate courses (**I.M2, II.M2, III.ME5**) for teachers with little present background in computing.

INSERVICE AND STAFF DEVELOPMENT PROGRAMS

In 1985, the NCTM showed concern for inservice and staff development programs by issuing a Position Statement. COMET strongly endorses this statement and reproduces it on page 6 with the permission of NCTM.

COMET also agrees with NCTM that excellent sources of help in implementing the ideas of their Position Statement are provided by the meetings, conferences, and publications of NCTM and its local, state, or provincial affiliated groups.

By **inservice** we mean programs for which college credit is not necessarily expected or offered. The focus is on a single topic or specific problem: for example, a new technology, certification requirements, teaching techniques, curriculum expectations, curriculum materials, educational research.

Teach a little well, rather than a lot badly.

To be most effective in inservice work, college and university departments of mathematics and of mathematics education must be involved with schools in ways such as

- Giving regular enrichment classes for school teachers and students.

- Assisting with teachers' professional days and conferences.

- Participating in student career days.

- Providing visiting speakers to schools.

- Organizing mathematics contests and competitions.

- Cooperating in parent-teacher activities.

- Sponsoring open houses.

Inservice and staff development experiences for teachers must fit the needs of teachers and of school programs. Participants expect information and help on specific problems with suggestions and materials that they can use in their classrooms. Since teacher participation in these programs is in addition to a full teaching load, the most successful programs provide opportunities for active involvement while making minimal demands on participants' outside time. Not all teachers are in the program because of a special interest in the announced topics; some may be there because their physical presence is required. The challenge of providing useful experiences to all participants requires consideration of at least six aspects of successful inservice programs.

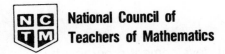
**National Council of
Teachers of Mathematics**

A Position Statement on . . .

Professional Development Programs for Teachers of Mathematics

Teachers of mathematics, like all professionals, require ongoing and cumulative professional development programs that enhance and maintain their teaching skills and knowledge. Because mathematics and education are disciplines that grow and change, teachers cannot depend on what they learned as undergraduates to carry them through their entire careers. Findings of research continually increase our understanding of teaching and learning. Further, social and technological changes increase the average citizen's need to understand and use mathematics. These forces demand reconsideration of the content and methods of mathematical instruction.

Curricular and instructional changes, however, do not occur automatically. The extent to which new ideas and techniques are integrated with current classroom practices depends on teachers' knowledge, motivation, and commitment to continued professional growth. The improvement of mathematics programs depends on well-prepared and well-informed teachers.

Such changes and improvement require teachers to have opportunities for high-quality professional development. The provision of these opportunities, which should maintain, enrich, and improve the skills and abilities that teachers need to serve their students best, is the shared responsibility of districts, schools, and individual teachers.

To help promote high-quality classroom instruction in mathematics, the National Council of Teachers of Mathematics encourages and supports the development and implementation of comprehensive professional development programs. The Council recommends that such programs be developed in accord with the following guidelines.

1. Professional development programs for teachers of mathematics should be based on a strong commitment to professional growth.
   a. An appropriate person should be responsible and accountable for the professional development of teachers.
   b. Sufficient time should be allocated for individuals to assess needs, plan activities, lead or participate in programs, and evaluate outcomes.

   c. Sufficient funds should be available to support professional development programs and ensure teachers' participation in them.

2. Professional development programs for teachers of mathematics should be carefully planned.
   a. Clear objectives should be established.
   b. The programs should improve students' learning experiences by improving the skills and knowledge of their teachers.
   c. Those whom the programs are designed to assist should contribute significantly in planning the programs.
   d. Extensive assessments of individual and collective needs should serve as bases of the programs.
   e. Current concerns and issues in mathematics education should be reflected in the content of the programs.
   f. The programs should be ongoing and cumulative.

3. Professional development programs for teachers of mathematics should recognize individual differences.
   a. Varied formats, including workshops, conferences, institutes, courses, and in-school discussion sessions, should be used.
   b. Programs should be tailored to meet the needs of teachers whose knowledge, skills, and experiences are diverse.

4. Professional development programs for teachers of mathematics should be effectively conducted and should include the following features:
   a. A blending of mathematical content and effective pedagogy
   b. Active participation of teachers
   c. Attention to the concrete, day-to-day problems of teachers.
   d. An integration of theory and practical applications
   e. Communication of objectives to participants
   f. Opportunities for teachers to practice new skills and techniques in the classroom
   g. Incorporation of support and follow-up activities.

5. Professional development programs for teachers of mathematics should be systematically evaluated, with attention to these issues:
   a. Determining whether the needs they are designed to meet have been satisfied
   b. Using the results from the evaluation to improve and develop future programs

(April 1985)

Important characteristics of a good inservice program include the following:

- **Publicity** should

    Be widespread.

    Give specific descriptions of the goals and intended activities.

    Tell participants what to expect.

- **Presenters** will often be a judicious mix of experienced teachers and supervisory personnel with college and university faculty, all of whom are

    Knowledgeable.

    Professional in their attitudes.

    Well prepared.

    Infectiously enthusiastic.

    Appreciative of the interests and background of the audience.

- **The environment** includes careful choice of

    Time of day; position in week. Spacing and number of regular (e.g., weekly) sessions.

    Place: in school or elsewhere; minimize participants' travel.

    Incentives for attendance: released or compensatory time, money, college credit, public recognition in local schools and newspapers, and letters to administrators and participants.

    Selection of audience: diverse enough to stimulate overall growth; homogeneous enough that all participants profit.

    Room, workshop, lecture hall. Availability of audio-visual aids.

    Avoidance of distraction: lighting, temperature, ventilation, seating, sound-proofing.

- **Content** is determined by

    Teachers' requests and needs.

    New technology.

Teaching techniques.

Research.

Curriculum development.

Enrichment.

Creation of awareness of human and material resources.

Recruitment for longer programs.

Review of previous knowledge.

Fine-tuning of existing skills.

- **The presentation**

    Is based on clear concepts and goals.

    Is well motivated.

    Concerns few topics in any one session.

    Uses a variety of media.

    Allows audience participation.

    Involves hands-on experience.

    Includes hand-outs of previously prepared material.

    Engenders group discussion.

    Provides a model for teachers' presentations, in their own classrooms.

    Gives participants ideas and materials for immediate use in the classroom.

- **Follow-up** includes

    Evaluation of program and its effectiveness.

    Administrative support for implementation.

    Feedback from participants to presenters.

    Subsequent meetings, visits, and courses.

In summary:

• Teachers of mathematics at all levels need to take an interest in what is going on at all other levels.

• Classroom teachers cannot avail themselves of programs which do not exist or whose existence is unknown to them.

• Colleges and universities have an important role to play and must approach schools to discover what inservice help is needed.

• Supervisors must consult with teachers, administrators, and places of higher education on the type of inservice programs needed and where and how to establish them.

• Everyone must take initiative. No one must merely hope that someone else is doing something about the problems.

References

Leigh Childs, *San Diego Math Network Training Guide*, San Diego County Office of Education, 1985.

Alan Osborne (ed.), *An In-Service Handbook for Mathematics Education*, NCTM, Reston, VA, 1977.

J. Price and J.D. Gawronski (eds.), *Changing School Mathematics*, NCTM, Reston, VA, 1981.

Ross Taylor, *Professional Development for Teachers of Mathematics: A Handbook*, NCTM, Reston, VA 1986.

MASTER'S DEGREE PROGRAMS

The main body of these GUIDELINES outlines master's programs for teachers of mathematics at three Levels. There are numerous course descriptions.

The Course descriptions are for SAMPLE courses.

We are not always as specific or as unambiguous as we would wish. Course titles, even detailed course descriptions, are subject to differing interpretations.

Find fresh material, or at least a fresh viewpoint.

Many presentations will begin with a review of prerequisite and background material, but do not merely repeat material that is standard in bachelor's programs for prospective mathematics teachers.

Courses suggested for one Level may also serve other Levels.

If it is necessary that a course serve teachers at more than one level, the presenter must keep this in mind. For example, it will often be expedient to combine courses for Level I and Level II teachers, but keep in mind the enormous range from Grade 1 to Grade 9. In these GUIDELINES notations such as **I.M5** and **III.ME2** are used to denote Level I mathematics course number 5 and Level III mathematics education course number 2, and so on.

Instructors must select the references most appropriate to the course they plan to teach.

We do not usually recommend texts. For some courses, good texts may still need to be written. We give references from a broad spectrum of literature:

- Books suitable for school mathematics classes.

- Enrichment material for school libraries.

- Undergraduate texts for mathematics education courses.

- Texts for mathematics education courses.

- Graduate texts for mathematics or mathematics education courses.

- Additional background material for master's program participants.

- Special references to challenge the expert.

10

Some references are out of print, but should be pursued in libraries. References often contain good bibliographies. An important part of education is to acquire the ability to search and appraise the literature. Both traditional and electronic searches can guide instructors and students to a broad and deep understanding of their subject. References mentioned for one course will often be appropriate to others.

For recent references, keep an eye on the MAA's *College Mathematics Journal* (Media Highlights section), *Mathematics Magazine,* and *American Mathematical Monthly*; NCTM's *Mathematics Teacher, Arithmetic Teacher,* (New Publications and Software sections) and *Journal of Research in Mathematics Education*; the Mathematical Association of Great Britain's *Mathematical Gazette*, and the journal *School Science and Mathematics.*

MASTER'S PROGRAMS WITH A MATHEMATICS CONCENTRATION FOR ELEMENTARY SCHOOL TEACHERS (LEVEL I)

These programs are designed to meet teachers' needs by increasing their knowledge in at least three different areas.

- Mathematical content.

- Pedagogical methods for teaching the content to children.

- Patterns in the mathematical development of children.

In the mathematics courses, special emphases are placed on the conceptual and intuitive development of the topics in a setting that encourages talking and writing about the topics.

The minimum mathematics background (9 semester hours) for full admission to a master's program comprises courses 1, 2, and 3 on pages 7-17 of the 1983 RECOMMENDATIONS:

1. Fundamental Mathematical Concepts I

2. Fundamental Mathematical Concepts II

3. Geometry for Elementary and Middle School Teachers

These courses, or their equivalents, are to be completed before the master's program courses are taken.

Master's programs that are based on sound undergraduate preparation and which contain 15 semester hours of mathematical content and mathematics education will enable Level I teachers to become effective mathematics specialists. Courses **I.M1 - I.M5** are sample mathematics courses and **I.ME1 - I.ME4** are sample mathematics education courses that could constitute an appropriate program. However, special effort should be made to fit courses to the backgrounds of the participants, even though it may be necessary to offer some of the same courses to both Level I and Level II teachers.

DESCRIPTIONS OF MATHEMATICS COURSES FOR LEVEL I

The mathematics content courses **I.M1 - I.M5** are intended to provide a background of knowledge beyond that expected to be taught in the elementary school. At the same time, it is important in each of these courses to discuss what *is* appropriate for the elementary school classroom and, where possible, to model the way in which it should be taught. These courses are for teachers. The goal is not only to broaden understanding but also to enhance teaching skills. Most class members will have had several years of teaching experience and much can be accomplished by asking them to share their expertise. In particular, they should complete projects and prepare sample lessons to be shared with other members of the class.

I.M1 Problem Solving in Elementary School Mathematics

Development of banks of problems appropriate to various levels and selected from informal geometry, measurement, number concepts, arithmetic, and logic; challenging enough to provoke interest, but realistic enough for successful experiences.

> What makes a good problem?
>
> Use of student generated problems.
>
> The art of problem posing. What if...?

Heuristics; problem-solving techniques; Pólya's stages of problem solving; specific strategies.

> Guess and check.
>
> Draw a diagram.
>
> Use models.
>
> Use both inductive and deductive reasoning.
>
> Consider simpler or related problems.
>
> Use simulation.
>
> Look back and reflect.

> Find a pattern.
>
> Make a list.
>
> Work backwards.

Pedagogical issues.

> Fitting problem solving into the curriculum.
>
> Classroom management; working in groups.
>
> Use of problem solving to develop new concepts and skills.
>
> Fostering positive attitudes toward problem-solving.
>
> Evaluation and testing.

References (see also II.M1 and I.ME4)

The Arithmetic Teacher, Teaching Problem Solving, NCTM, Reston, VA, Feb. 1982.

Alma Bingham, *Improving Children's Facility in Problem Solving*, Columbia University Press, New York, 1958.

Maxey Brooke, *Coin Games and Puzzles*, Dover, New York, 1963.

Marilyn Burns, *The Book of Think*, Little-Brown and Co., Boston, 1976.

Marilyn Burns, *The Good Time Math Event Book*, Creative Publications, Palo Alto, CA, 1977.

Randall Charles and Frank Lester, *Teaching Problem Solving: What, Why and How*, Dale Seymour, Palo Alto, CA, 1982.

Randall Charles, Frank Lester, and Phares O'Daffer, *How to Evaluate Progress in Problem Solving*, NCTM, Reston, VA, 1987.

Eric Doubleday, *Test Your Wits*, Vol. 5, Ace Books, New York, 1962.

Jack Frolichstein, *Mathematical Fun, Games and Puzzles*, Dover, New York, 1962.

James G. Greeno, A study in problem solving, in R. Glaser (ed.), *Advances in Instructional Psychology*, Vol. 1, Lawrence Erlbaum, Hillsdale, NJ, 1983.

James G. Greeno, Forms of understanding in mathematical problem solving, in S.G. Paris, G.M. Olson and H.W. Stevenson (eds.), *Learning and Motivation in the Classroom*, Lawrence Erlbaum, Hillsdale, NJ, 1983.

Anita Harnadek, *Critical Thinking*, Midwest Publications, Troy, MI, 1976.

Ed Labinowicz, *Learning from Children: New Beginnings for Teaching Numerical Thinking*, Addison-Wesley, Reading, MA, 1985.

Steven P. Meiring, *Problem Solving: A Basic Mathematics Goal*, Vols. 1 and 2, Ohio Department of Education, 1981.

Doyal Nelson and Joan Worth, *How to Choose and Create Good Problems for Primary Children*, NCTM, Reston, VA, 1983.

Ronald C. Read, *Tangrams--330 Puzzles*, Dover, New York, 1965.

Barbara Reys, *Elementary School Mathematics: What Parents Should Know about Problem Solving*, NCTM, Reston, VA, 1982.

Marilyn N. Suydam and J. Fred Weaver, Research on problem solving: Implications for elementary school classrooms, *The Arithmetic Teacher*, 25 (1977) 40-42.

John W. Wilson, What skills build problem-solving power?, *The Instructor*, 76 (1967) 79-80.

I.M2 Calculators and Computers in Elementary School Mathematics

To help elementary school teachers to become adept at using calculators and computers to enhance instruction in mathematics.

Use of calculators -- what is appropriate and what is not.

Calculator operation -- keying sequences and order of operations, use of memory and constant functions, use of arithmetic function keys.

Calculator applications -- estimation and approximation, pattern development and recognition, guess and check methods, computations too large for the display, and integer division with remainder.

Use of computers -- what is appropriate and what is not.

Computer graphics -- use of Logo and turtle geometry to explore geometric concepts and relationships and to develop higher order thinking skills.

Identification of exemplary software.

Problem solving with calculators and computers.

References (see also **II.M2 and III.ME5**)

The Arithmetic Teacher, Calculators, NCTM, Reston, VA, Feb. 1987.

Donna Beardon, *1,2,3, My Computer and Me!*, Reston Publications, Englewood Cliffs, NJ, 1980.

Donna Beardon, Kathleen Martin, and Jim Muller, *The Turtle Source Book*, Dale Seymour, Palo Alto, CA.

Rick Billstein, Shlomo Liebskind and Johnny W. Lott, *Apple Logo: Programming and Problem Solving*, Benjamin/Cummings, Menlo Park, CA, 1986.

Joanne Capper, Computers in Education, *Research Into Practice Digest*, 1 (1986) No. 3.

J.P. East, Computer Education for Elementary Schools: A Course for Teachers, in *Topics: Computer Education for Colleges of Education*, Association for Computing Machinery, New York, 1983.

Janet Morris, *How to Develop Problem Solving Using a Calculator*, NCTM, Reston, VA, 1981.

Seymour Papert, *Mindstorms: Children, Computers, and Powerful Ideas*, Basic Books, New York, 1980.

Donald Watt and Molly Watt, *Teaching With Logo*, Addison-Wesley, Menlo Park, CA, 1986.

Courseware

Appleworks, Clarion Computer Company, Cupertino, CA.

Creative Play, Lawrence Hall of Science, University of California, Berkeley, CA.

EduCalc II, Grolier Electronic Publishing, Danbury, CT.

Exploring Tables and Graphs I, Weekly Reader Family Software, Middletown, CT.

EZ Logo, MECC, St. Paul, MN.

Factory, Sunburst Communications, Pleasantville, NY.

Gertrude's Puzzles, Sunburst Communications, Pleasantville, NY.

Problem Solving Strategies, Minnesota Educational Computing Consortium, St. Paul, MN.

Teasers by Tobbs, Sunburst Communications, Pleasantville, NY.

Terrapin Logo, Terrapin Inc., Cambridge, MA.

I.M3 Concepts of Geometry

Extension of techniques for visualization and recognition of two-and three-dimensional geometric figures.

Pedagogical uses of Logo turtle graphics.

Interrelations among measurement and numerical ideas and skills.

Development of methods for teaching geometric concepts at different grade levels.

References

H. Abelson and Andrea diSessa, *Turtle Geometry*, M.I.T. Press, Cambridge, MA, 1980.

Patricia S. Davidson and Robert E. Willcut, *Spatial Problem Solving with Cuisenaire Rods*, Cuisenaire Company of America, New Rochelle, NY, 1983.

D. Hilbert and S. Cohn-Vossen, *Geometry and the Imagination*, P. Neményi (trans.), Chelsea, New York, 1952.

Alan Hoffer (ed.), *Geometry and Visualization*, Mathematics Resource Project, University of Oregon, Eugene, OR, 1977.

P.G. O'Daffer and S.R. Clemens, *Geometry: An Investigative Approach*, Addison-Wesley, Reading, MA, 1977.

Seymour Papert, *Mindstorms: Children, Computers, and Powerful Ideas*, Basic Books, New York, 1980.

Janet Shroyer and William Fitzgerald, *Mouse and Elephant: Measuring Growth*, Addison-Wesley, Menlo Park, CA, 1986.

David Eugene Smith, *The Teaching of Geometry*, NCTM, Reston, VA.

Mary Jean Winter, *et al*, *Spatial Visualization*, Addison-Wesley, Menlo Park, CA, 1986.

Grace Chisholm Young and W.H. Young, *Beginner's Book of Geometry*, Chelsea, New York, 1970.

IM.4 Statistics and Probability in The Elementary School Curriculum

Introduction to data analysis.

The place of data analysis in modern statistics.

Techniques of data analysis including mean, median, standard deviation, interquartile range, histogram, stem-and-leaf plots, box-and-whisker plots, scatter plots, median fit.

Probability.

Empirical probability.

Simulation, including the use of random number tables and computer random number generators.

Theoretical probability including sample spaces, and introduction to counting (fundamental principle, permutations, combinations), the addition rule, conditional probability, the multiplication rule.

Introduction to sample surveys and confidence intervals.

Sampling including random sampling and bias in sampling.

Well-known polls such as Gallup, Nielsen, Current Population Survey.

Confidence intervals for percentages in polls.

Reference (See also II.M4)

Mrudrulla Gnanadesikan, Richard L. Scheaffer, and Jim Swift, *The Art and Techniques of Simulation*, Dale Seymour, Palo Alto, CA, 1986.

James M. Landwehr, Jim Swift, and Ann E. Watkins, *Exploring Surveys and Information from Samples*, Dale Seymour, Palo Alto, 1986.

James M. Landwehr and Ann E. Watkins, *Exploring Data*, Dale Seymour, Palo Alto, CA, 1986.

Claire Newman, Tom Obremski and Richard Scheaffer, *Exploring Probability*, Dale Seymour, Palo Alto, CA, 1986.

Ivan Niven, *Mathematics of Choice*, New Mathematical Library 16, MAA, Washington DC, 1965.

Elizabeth Phillips, *et al*, *Probability*, Addison-Wesley, Menlo Park, CA, 1986.

Stuart Choate and Albert Shulte, *What are my Chances*, Creative Publications, Palo Alto, CA.

Albert P. Shulte and James R. Smart (eds.), *Teaching Statistics and Probability*, 1981 Yearbook, NCTM, Reston, VA.

I.M5 Numbers: An Indepth Look

Division. The Euclidean algorithm. The fundamental theorem of arithmetic. Number systems in which unique factorization does not hold.

Divisibility rules in base 10 and other bases, and their proofs.

Equivalence of fractions and its use to perform the four arithmetic operations.

Representations of rational numbers as repeating decimals in base 10 and other bases. Is $0.\overline{9} = 1$?

Proof of the irrationality of $\sqrt{2}$ by geometry and by algebra.

Golden section and the golden mean.

Ordinals and cardinals. Where does zero fit in? Infinite sets. Are there more rational numbers than integers? Are there more real numbers than integers? Are there more real numbers than rational numbers? The continuum hypothesis.

Peano's axioms used to prove simple theorems. Comparison of Peano's axioms with von Neumann's set theory approach.

Complex numbers. The fundamental theorem of algebra.

References

Asger Aaboe, *Episodes from the Early History of Mathematics*, New Mathematics Library 13, MAA, Washington DC, 1964.

The Arithmetic Teacher, Rational Numbers, NCTM, Reston, VA, Feb. 1984.

Douglas M. Campbell, *The Whole Craft of Number*, Prindle, Weber, and Schmidt, Boston, 1976.

John H. Conway, *On Numbers and Games*, Academic Press, London and New York, 1976.

John H. Conway and Richard Guy, *The Book of Numbers*, W.H. Freeman, New York, 1988.

Tobias Dantzig, *Number, The Language of Science*, 4th edition, Doubleday, New York, 1856.

Philip J. Davis, *The Lore of Large Numbers*, New Mathematical Library 6, MAA, Washington DC, 1961.

Donald Knuth, *Surreal Numbers*, Addison-Wesley, Reading, MA, 1974.

Frank Land, *The Language of Mathematics*, John Murray, London, 1975.

Ivan Niven, *Numbers: Rational and Irrational*, New Mathematical Library 1, MAA, Washington DC, 1961.

Oystein Ore, *Invitation to Number Theory*, New Mathematical Library 20, MAA, Washington DC, 1967.

Peter Roza, *Playing with Infinity*, Mathematical Explorations and Excursions, Dover, NY, 1961.

D.E. Smith and J. Ginsburg, *Numbers and Numerals*, NCTM, Reston, VA, 1937.

Sherman K. Stein, *Mathematics, the Man-Made Universe*, 3rd edition, W.H. Freeman, San Francisco, 1976.

Leo Zippin, *Uses of Infinity*, New Mathematical Library 7, MAA, Washington DC, 1962.

DESCRIPTIONS OF MATHEMATICS EDUCATION COURSES FOR LEVEL I
(Read also the introductory statement "Mathematics Education Courses for Level III.")

I.ME1 Teaching Strategies in Elementary School Mathematics

This course emphasizes and extends several principles of instruction. Children learn better with:

Carefully crafted lessons emphasizing concept development over drill.

The use of manipulative materials that accurately represent or model mathematical concepts.

Active involvement in mathematics.

New or extended concepts presented explicitly in relation to previously learned material.

Motivation provided by enrichment materials.

Opportunities to talk and read mathematics.

Use of calculators as well as paper and pencil.

Instruction based on the changing characteristics of their thinking.

New mathematical concepts and skills developed from problems.

The active involvement of their parents.

The strategies of teaching this course must exemplify these principles. It is the responsibility of the instructor to plan activities that strengthen the participants' commitment to, and extend their capabilities to use, these principles.

The course includes:

Analysis of relations between learning theories and methods of classroom instruction.

Comparison and evaluation of methods of classroom management and structure, laboratory approach, individualized mathematics programs, and computer assisted instruction.

The roles of grouping, pacing, and evaluation.

Strategies for concept development, learning basic facts, and molding positive attitudes towards mathematics.

Strategies for parent involvement.

In order to determine the content of the course and plan its presentation, instructors must know local needs, the perceptions of teachers, and the strengths and weaknesses of the local school mathematics program (National Assessment results, local test performances). Examples and activities must be selected with emphasis on topics which are not well or comfortably taught in schools, and may include:

Sets, attributes, sorting, matching, counting, numerousness, numerals, comparing, ordering, generalizing.

Number / numeral patterns, equivalence, use of calculators.

Geometric designs, geometry in the environment, lines polygons, symmetry, spatial visualization.

Deductive thinking, organization of information, solution strategies.

References

Mary Baratta-Lorton, *Work Jobs II, Number Activities for Early Childhood*, Addison-Wesley, Menlo Park, CA 1976.

Mary Baratta-Lorton, *Mathematics Their Way*, Addison-Wesley, Menlo Park, CA 1976.

Gary G. Bitter, Mary Hatfield, and Nancy Edwards, *Elementary and Middle School Mathematics: A Comprehensive Approach*, Allyn and Bacon, Needham Heights, NJ, 1988.

Grace M. Burton, *Toward a Good Beginning: Teaching Early Childhood Mathematics*, Addison-Wesley, Menlo Park, CA, 1985.

Randall I. Charles, *et al, Problem-Solving Experiences in Mathematics: Grades 1,2,3,...,8*, Addison-Wesley, Menlo Park, CA, 1985.

Mark J. Driscoll, *Research Within Reach: Elementary School Mathematics*, NCTM, Reston, VA, 1981.

W. George Cathcart (ed.), *The Mathematics Laboratory: Readings from The Arithmetic Teacher*, NCTM, Reston, VA, 1977.

James W. Heddens, *Today's Mathematics*, Science Research Associates, Chicago, 1984.

George Immerzeel and Melvin L. Thomas (eds.), *Ideas from The Arithmetic Teacher: Grades 1-4 Primary*, NCTM, Reston, VA, 1982.

Eugene Krause, *Mathematics for Elementary Teachers: A Balanced Approach*, Heath, Lexington, MA, 1987.

Joseph Payne, *et al, Mathematics Learning in Early Childhood*, 1975 Yearbook, NCTM, Reston, VA.

Thomas R. Post (ed.), *Teaching Mathematics in Grades K-8: Research Based Methods*, Allyn and Bacon, Boston, 1988.

Barbara Rey, *Elementary School Mathematics: What Parents Should Know About Estimation*, NCTM, Reston, VA, 1982.

Seaton E. Smith and Carl A. Bachman (eds.), *Games and Puzzles for Elementary and Middle School Mathematics: Readings From The Arithmetic Teacher*, NCTM, Reston, VA, 1975.

Edith L. Somervell, *A Rhythmic Approach to Mathematics*, NCTM, Reston, VA, 1975.

Larry Sowder, *Didactics and Mathematics: The Art and Science of Learning and Teaching Mathematics*, Creative Publications, Palo Alto, CA, 1978.

Marilyn N. Suydam and J. Fred Weaver, *Research: A Key to Elementary School Mathematics, NCTM, Reston, VA, 1975.*

Jean Kerr Stenmark, Virginia Thompson and Ruth Cossey, *Family Math*, Lawrence Hall of Science, University of California, Berkeley, CA, 1986.

Catherine Stern and M.B. Stern, *Children Discover Arithmetic*, 2nd edition, George C. Harrap, London, 1971.

Andria I. Troutman and Betty K. Lichtenberg, *Mathematics: A Good Beginning*, 3rd edition, Brooks Cole, Monterey, CA, 1987.

I.ME2 Current Issues and Research in Elementary School Mathematics

A course to help the participants make better use of the results of research in teaching and to recognize classroom teaching as an appropriate setting for experimentation and problem solving.

A thorough review of the literature on experimental and exemplary programs, results of recent national, state, and local assessments of student achievement in mathematics, and current research related to:

Concepts of number.

Skills with arithmetic operations.

Concepts of geometry in the plane and in space.

Concepts of and skills with measurement.

The notion of chance.

Problem solving and critical thinking.

The relation of these studies to local teaching practices and programs, instructional materials and evidence of local student achievement.

The effectiveness of standardized tests in measuring the attainment of program goals. Special attention to recent research findings concerning:

Curriculum structure.

Teaching approaches.

Factors affecting learning.

The use of instructional and enrichment materials.

The role of technology.

Impact of parent involvement.

Each participant identifies a significant problem area in teaching mathematics, formulates a plan to solve the problem, and evaluates the results of implementing the plan. To assist in the design of individual projects, the instructor must be familiar with local school mathematics programs and know how to design realistic research projects. The participant assumes a professional attitude devoted to improving what happens in the classroom.

References

Gary G. Bitter, Educational Technology and the future of mathematics education, *School Science and Mathematics*, 87 (1987) 454-465.

Joanne Capper, *Mathematical Problem Solving: Research Review and Instructional Implications*, Center for Research Into Practice, Washington DC, 1984.

Joanne Capper, *Computers in Mathematics Learning*, Center for Research Into Practice, Washington DC, 1986.

Elizabeth Fennema (ed.), *Mathematics Education: Research Implications for the 80's*, Association for Supervision and Curriculum Development, Alexandria, VA and NCTM, Reston, VA, 1981.

Chester E. Finn (ed.), *What Works: Research about Teaching and Learning*, U.S. Department of Education, Washington DC, 1986 (an answer to *A Nation at Risk*).

Journal for Research in Mathematics Education, NCTM, Reston, VA.

Mary Montgomery Lindquist (ed.), *Selected Issues in Mathematics Education*, McCutchan, Berkeley, CA, 1980.

School Science and Mathematics, School Science and Mathematics, Bowling Green, OH.

I.ME3 Diagnosis and Remediation in Elementary School Mathematics

Sequencing of mathematical knowledge.

Piaget, Bruner, Gagné, van Hiele, Hart-Kucheman-Booth.

Diagnosis of difficulties in

Basic concepts.

Associated vocabulary.

Classification.

Sets and subsets.

Size.

Counting.

Seriation in time and space.

Operations with whole numbers.

Beginning fractions and decimals.

Spatial relations.

Measurement.

Gathering information including

Written tests versus interviews.

Building a test or interview for diagnosis.

Appraisal of available commercial tests.

Detecting and correcting error patterns including

Undue identification of numbers with specific configurations.

Use of wrong operations.

Language related errors.

Use of wrong operands.

Wrong order of steps in an algorithm.

Wrong concept: e.g. "triangle" identified with special case (equilateral, isosceles, right-angle).

Random responses.

Designing remediation activities including

Self-appraisal.

Clear goals.

Use, and abandonment, of crutches.

Cultivation of ability to estimate.

Strengthening of chldren's self image.

Practice leading to immediate feedback.

Use of a variety of instructional procedures.

Children using colloquial language.

Small steps.

Using alternative algorithms.

Encouraging organization.

References

Robert B. Ashlock, *Error Patterns in Computation: A Semi-programmed Approach*, 2nd edition, Charles E. Merrill, Columbus, OH, 1976.

Robert B. Ashlock, Martin L. Johnson, John L. Wilson and Wilmer L. Jones, *Guiding Each Child's Learning of Mathematics: A Diagnostic Approach to Instruction*, Charles E. Merrill, Columbus, OH, 1983.

Gary G. Bitter, Jon M. Englehardt, and James Wiebe, *One Step at a Time: A Diagnostic/Prescriptive Mathematics Program*, Educational Materials Corporation, St. Paul, MN, 1977.

Jon M. Englehardt, Feedback in diagnostic testing, in *Clinical Investigations in Mathematics Education*, 1977 Annual Proceedings, Research Council for Diagnostic and Prescriptive Mathematics, Bowling Green, OH.

Frederika K. Reisman, *Diagnostic Teaching of Elementary School Mathematics: Methods and Content*, Rand McNally, Chicago, 1977.

I.ME4   The Psychology of Learning Mathematics

Connections between general psychological theories and the teaching of mathematics. Identification of several examples from the classroom for each of these domains of cognitive function:

Concept development.

Memory.

Readiness for learning.

Learning styles.

Problem solving.

Skill learning.

Information processing.

Perception.

Concept development.

The emphasis is on higher order skills (cognition and thinking) rather than on motor learning and other lower-level skills. The course includes frequent demonstrations of the relation between psychological principles and instructional decisions that mathematics teachers must make.

References   (see also II.ME4)

The Arithmetic Teacher, Mathematical Thinking, Feb. 1985, NCTM, Reston, VA.

T.P. Carpenter, Learning to add and subtract: An exercise in problem solving, in E.A. Silver (ed.), *Teaching and Learning Mathematical Problem Solving: Multiple Research Perspectives*, Lawrence Erlbaum, Hillsdale, NJ, 1985.

T.P. Carpenter and J.M. Moser, The development of addition and subtraction problem-solving skills, in T.P. Carpenter, J.M. Moser and T.A. Romberg (eds.), *Addition and Subtraction: A Cognitive Perspective*, Lawrence Erlbaum, Hillsdale, NJ, 1983.

T.P. Carpenter and J.M. Moser, The acquisition of addition and subtraction concepts, in R. Lesh and M. Landau (eds.), *Acquisition of Mathematics Concepts and Processes*, Academic Press, New York, 1983.

T.P. Carpenter and J. M. Moser, The acquisition of addition and subtraction concepts in grades one through three, *Journal for Research in Mathematics Education*, 15 (1984) 179-202.

Robert Case, General development influences on the acquisition of elementary concepts and algorithms in arithmetic, in T.P. Carpenter, J.M. Moser, and T.A. Romberg (eds.), *Addition and Subtraction: A Cognitive Perspective*, Lawrence Erlbaum, Hillsdale, NJ, 1983.

Richard W. Copeland, *How Children Learn Mathematics: Teaching Implications of Piaget's Research*, 4th edition, Macmillan, New York, 1984.

Ann Floyd (ed.), *Developing Mathematical Thinking*, Addison-Wesley, Reading, MA, 1982.

Rachel Gelman and C.R. Gallistel, *The Child's Understanding of Number*, Harvard University Press, Cambridge, MA, 1978.

James G. Greeno, Conceptual entities, in D. Gentner and A.L. Stevens (eds.), *Mental Models*, Lawrence Erlbaum, Hillsdale, NJ, 1983.

Constance Kamii, *Number in Preschool and Kindergarten: Educational Implications of Piaget's Theory*, National Association for the Education of Young Children, Washington DC, 1982.

Kathleen M. Hart, *Ratio: Children's Strategies and Errors*, NFER-Nelson, Windsor, UK, 1984.

Ed Labinowicz, *The Piaget Primer: Thinking, Learning, Teaching*, Addison-Wesley, Menlo Park, CA, 1980.

Richard Meyer, *Thinking, Problem Solving, Cognition*, W.H. Freeman, Boston, 1983.

J.N. Payne (ed.), *Mathematics Learning in Early Childhood*, NCTM, Reston, VA, 1975.

L. Resnick and W.W. Ford, *The Psychology of Mathematics Instruction*, Lawrence Erlbaum, Hillsdale, NJ, 1981.

M.F. Rosskopf (ed.), *Children's Mathematical Concepts: Six Piagetian Studies in Mathematical Education*, Teachers College Press, Columbia University, New York, 1975.

Leslie P. Steffe, *Research on Mathematical Thinking of Young Children*, NCTM, Reston, VA, 1975.

MASTER'S PROGRAMS FOR TEACHERS OF MIDDLE SCHOOL MATHEMATICS (LEVEL II)

These programs are designed to

- Deepen and extend the teacher's knowledge of mathematics.

- Enable teachers to foster and maintain interest in mathematics among students who have rapidly changing levels of maturity.

- Provide enrichment sufficient to interest and challenge all students, including the mathematically gifted.

- Facilitate diagnosis of weaknesses and learning problems.

- Suggest instruction appropriate to sensitive adolescents.

A good balance of courses from mathematics and mathematics education is essential to the development of classroom specialists who are prepared in content and also are able to communicate it to students of widely varying abilities and backgrounds in grades five through nine.

A master's program for middle school teachers requires at least 30 semester hours, including at least

- 6 hours in mathematics education (instructional strategies, diagnosis and prescription, current issues and research).

- 15 hours of courses providing enrichment of middle school mathematics topics (problem solving, calculators and computers, probability and statistics, informal geometry, elementary algebra).

- 9 hours of mathematics, chosen to refresh and extend the teacher's own knowledge of mathematics (see next three paragraphs).

A teacher entering a master's program should have completed the Level II program described in the 1983 RECOMMENDATIONS which includes a choice of four of the following Level III courses, the starred courses being particularly appropriate:

* Discrete Mathematics

 Introduction to Computing

* Mathematics Appreciation

 Linear Algebra

* Probability & Statistics

* Number Theory

 Geometry

 Abstract Algebra

* History of Mathematics

 Mathematical Modelling & Applications

The mathematics content courses are chosen to complement the prerequisites by filling gaps in the teacher's background and extending knowledge in areas of particular interest. When designing the mathematics content part of the program, consult the descriptions in the 1983 RECOMMENDATIONS of the ten courses listed above and also consult courses **III.M1** through **III.M15** of these Guidelines.

If teachers are interested in a master's program, but do not have preparation comparable to the Level II recommendations, a program should be designed which incorporates the Level II recommendations. This is particularly important, for example, in states where there is no specific middle school certification or where teachers with only Level I preparation may wish to develop a speciality in teaching middle school mathematics.

Sample courses in mathematics, **II.M1 - II.M6**, and in mathematics education, **II.ME1 - II.ME4**, suitable for a Level II masters program, are described below.

DESCRIPTIONS OF MATHEMATICS COURSES FOR LEVEL II

II.M1 Problem Solving in Middle School Mathematics

See the course description for **I.M1**. Here, in **II.M1**, the problems are selected from probability, statistics, and elementary algebra. Pedagogical issues will include:

Extension and evaluation of skills in the teaching of problem solving in the middle school.

Evaluating the growth of problem-solving skills of students.

References

Bonnie Averbach and Orin Chein, *Problem Solving Through Recreational Mathematics*, W.H. Freeman, San Francisco, 1980.

Lesley R. Booth, Difficulties in algebra, *The Australian Mathematics Teacher*, 42 (1986) 2-4.

Jack Briggs, Franklin D. Demana, and Alan Osborne, Moving into algebra: Developing the concepts of variable and function, *The Australian Mathematics Teacher*, 42 (1986) 5-8.

Stephen Brown, *et al, Creative Problem Solving*, Bureau of Mathematics Education, Albany, NY, 1987.

Stephen Brown and Marion Walter, *The Art of Problem Posing*, Dale Seymour, Palo Alto, CA, 1983.

William A. Brownell, Problem Solving, in *The Psychology of Learning*, Part II, pp. 415-443, National Society for the Study of Education, 1942.

Marilyn Burns, *The I Hate Mathematics! Book*, Little-Brown and Co., Boston, 1975.

Franklin D. Demana, Joan Leitzel, and Alan Osborne, *Getting Ready for Algebra*, D.C. Heath, Lexington, MA, 1987.

Daniel T. Dolan and James Williamson, *Teaching Problem-Solving Strategies*, Addison-Wesley, Menlo Park, CA, 1983.

Henry E. Dudeney, *Amusements in Mathematics*, Dover, New York, 1958.

Henry E. Dudeney, *536 Puzzles and Curious Problems*, Charles Scribner's Sons, New York, 1967.

William M. Fitzgerald, *Proceedings of the Honors Teachers Workshop of Middle Grade Mathematics*, Department of Mathematics, Michigan State University, East Lansing, MI, 1984.

Martin Gardner, *Scientific American Book of Mathematical Puzzles and Diversions*, Simon & Schuster, New York, 1959.

Martin Gardner, *2nd Scientific American Book of Mathematical Puzzles and Diversions*, Simon & Schuster, New York, 1961.

Martin Gardner and Robert Tappey, *The Paradox Box*, *Scientific American*, W.H. Freeman, San Francisco, 1975.

Carole Greenes, John Gregory and Dale Seymour, *Successful Problem Solving Techniques*, Creative Publications, Palo Alto, CA, 1979.

Anita Harnadeck, *Mathematical Reasoning*, Midwest Publications, Troy, MI, 1969.

Boris A. Kordensky, *The Moscow Puzzles: 359 Mathematical Recreations*, Charles Scribner's Sons, New York, 1972.

Stephen Krulik and Robert E. Reys (eds.), *Problem Solving in School Mathematics*, 1980 Yearbook, NCTM, Reston, VA.

L.H. Longley-Cook, *Fun with Brain Puzzlers*, Fawcett Publications, Greenwich, CT, 1965.

Sam Loyd, *Mathematical Puzzles of Sam Loyd*, Dover, New York, 1959.

John Mason with Leon Burton and Kaye Stacey, *Thinking Mathematically*, Addison-Wesley, Reading, MA, 1982.

Sid Rachlin (ed.), *Problem Solving in the Mathematics Classroom*, NCTM, Reston, VA, 1982.

Thomas Romberg, Individual Differences and The Common Curriculum, in *The Common Curriculum in Mathematics*, 82nd Yearbook, Part I, National Society for the Study of Education, University of Chicago Press, Chicago, 1983.

Dale Seymour and Margaret Shedd, *Finite Differences: A Pattern-Discovery Approach to Problem Solving*, Dale Seymour, Palo Alto, CA, 1973.

Gwen Shufelt (ed.), *An Agenda for Action*, 1983 Yearbook, NCTM, Reston, VA, 1982.

Linda Silvey and James R. Smart (eds.), *Mathematics for the Middle Grades (5-9)*, 1982 Yearbook, NCTM, Reston, VA.

Marion I. Walter, *Boxes, Squares, and Other Things*, NCTM, Reston, VA, 1970.

Grayson H. Wheatley, The right hemisphere's role in problem solving, *The Arithmetic Teacher*, 25 (Nov., 1977) 36-38.

II.M2 Calculators and Computers in Middle School Mathematics

To help middle school or junior high school teachers become adept at using calculators and computers to enhance instruction in mathematics. An integral part of this course is learning to distinguish situations in which calculators or computers are appropriate from situations where paper and pencil and mental calculation are more appropriate. Hopefully, participants will have had the equivalent of:

> Course 4. Algebra and Computing for Elementary and Middle School
> Teaching

described on pages 19 through 21 of the 1983 RECOMMENDATIONS. Topics include:

Use of calculators -- what is appropriate, what is not.

Calculator operation -- keying sequences and hierarchy of operations, use of memory, constant function, special function keys.

Calculator application -- estimation and approximation, pattern development and recognition, guess and check methods, computations too large for the display, integer division with remainders.

Use of computers -- what is appropriate and what is not.

Computer graphics -- use of Logo and turtle geometry to explore geometric concepts and relationships and to develop higher order thinking skills.

Use of computers to investigate ideas in pre-algebra, number theory, probability and statistics.

Use of simulation and problem solving software in exploring mathematics.

Identification of exemplary software, both good and bad, for each grade level.

Effective use of software in instruction.

Problem solving with calculators and computers.

References (see also I.M2 and III.ME5)

Irving Adler, *Thinking Machines: A Layman's Introduction to Logic, Boolean Algebra and Computers*, NCTM, Reston, VA, 1975.

Stanley J. Bezuszka and Margaret J. Kenney, *Number Treasury: A Sourcebook of Problems for Calculators and Computers*, Dale Seymour, Palo Alto, CA, 1982.

Gary G. Bitter (with R. Camuse), *Using a Microcomputer in the Classroom*, Prentice Hall, Englewood Cliffs, NJ, 1988.

Gary G. Bitter and J. Mikesell, *Handbook for Teaching with the Hand Held Calculator*, Allyn and Bacon, Boston, 1988.

Gary G. Bitter, Educational Technology and The Future of Mathematics Education, in *School Science and Mathematics*, 87 (1987) 454-465.

Bruce C. Burt (ed.), *Calculators: Readings from The Arithmetic Teacher and The Mathematics Teacher*, NCTM, Reston, VA, 1979.

Joanne Capper, *Computers in Mathematics Learning*, Center for Research Into Practice, Washington, DC.

Viggo P. Hansen and Marilyn J. Zweng (eds.), *Computers in Mathematics Education*, 1984 Yearbook, NCTM, Reston, VA.

Rochelle Heller, *et al*, *Logo Worlds*, Computer Science Press, Rochelle, MD, 1985.

Beverly Hunter, *My Students Use Computers: Computer Literacy in the K-8 Curriculum*, Dale Seymour, Palo Alto, CA, 1984.

International Commission on Mathematical Instruction, *The Influence of Computers and Informatics on Mathematics and its Teaching*, Strassbourg Conference, 1985, MAA, Washington DC, 1986.

Glen Kleiman, *Brave New Schools: How Computers Can Change Education*, Dale Seymour, Palo Alto, CA, 1984.

Arthur Luehrman and H. Peckman, *Appleworks Spreadsheets*, Computer Literacy Press, Gilroy, CA, 1987.

David Moursund, *Introduction to Computers in Education for Elementary and Middle School Teachers*, International Council for Computers in Education, Eugene, OR, 1981.

Richard J. Shumway, Calculators and Computers, in Thomas R. Post (ed.), *Teaching Mathematics in Grades K-8: Research Based Methods*, Allyn and Bacon, Boston, 1988.

J. Poirot, R. Taylor and J. Powell, *Teacher Education Topics: Computer Education for Elementary and Secondary Schools*, Association for Computing Machinery, New York, 1981.

Robert P. Taylor (ed.), *The Computer in the School: Tutor, Tool, Tutee*, Teachers College Press, Columbia University, New York, 1980.

David Thornberg, *Beyond Turtle Graphics*, Addisoon-Wesley, Menlo Park, CA, 1986.

David Williams, *Mathematics Teachers Complete Calculator Handbook*, Prentice Hall, Englewood Cliffs, NJ, 1984.

II.M3 Concepts of Geometry in the Middle School

A previous college geometry course is assumed: for example,

Course 3: Geometry for Elementary and Middle School Teachers

on pages 16 through 19 of the 1983 RECOMMENDATIONS. Topics include:

A study of geometric objects (lines, planes, polygons, polyhedra, circles, spheres, cylinders) and their relationships (congruence, symmetry, similarity) through a variety of different approaches:

Intuitive (constructions, geoboards, paper folding).

Coordinate (analytic; the relation between algebra and geometry).

Vector (position, motion, transformation).

Synthetic (the role of axioms).

A comparison of these approaches and their appropriateness in various situations. The existence and visualization of other geometries (spherical, projective, affine, hyperbolic) are explored by models and sample theorems.

The relationship of geometry in grades 5-9 to geometry in later grades.

The course builds from an informal intuitive approach, stressing visualization in two and three dimensions, to the development of skills in mathematical reasoning in geometry.

References (see also **I.M3, III.M2,** and **III.M3**)

Tom Banchoff/Paul Strauss Production, *Hypercube*, Brown University, Providence, RI, 1978.

Martha Boles and Rochelle Newman, *The Golden Relationship: Art, Math, Nature, Book I: Universal Patterns*, Pythagorean Press, Bradford, MA, 1983.

Kenneth B. Henderson (ed.), *Geometry in the Mathematics Curriculum*, 36th Yearbook, NCTM, Reston, VA, 1973.

Alan Hoffer, *Geometry*, Addison-Wesley, Reading, MA, 1979.

Gerald Jenkins and Anne Wild, *Mathematical Curiosities*, 3 booklets, Park West, NY, 1980.

Phares B. O'Daffer and Stanely R. Clemens, *Geometry: An Investigative Approach*, Addison-Wesley, Reading, MA, 1977.

Report on the Teaching of Geometry, Mathematics Association, Leicester, UK, 1951.

Michael Serra, *Inductive Geometry: A Discovery Approach*, Peter Rasmussen, San Francisco, 1988.

Richard J. Schumway, Calculator and computers, in *Teaching Mathematics in Grades K-8: Research Based Methods*, Thomas R. Post (ed.), Allyn and Bacon, Boston, 1988, 410-437.

Marion Walter, *Make A Bigger Puddle; Make a Smaller Worm*, Deutsch, 1971.

Marion Walter, *Another, Another, Another and More*, Deutsch, 1975.

Marion Walter, *The Mirror Puzzle Book*, Park West, NY, 1985.

Courseware

Fred Daly, Bob Burn and Chris Forecast, *Tesselations*, Cambridge University Press, London, 1985.

II.M4 Statistics and Probability in the Middle School

Introduction to data analysis.

    The place of data analysis in modern statistics.

    Techniques of data analysis including mean, median, standard deviation, interquartile range, histogram, stem-and-leaf plots, box-and-whisker plots, scatter plots, correlation, median fit, regression lines.

Probability.

    Empirical probability.

    Simulation, including the use of random number tables and computer random number generators.

    Theoretical probability including sample spaces.

Introduction to sample surveys and confidence intervals.

    Sampling including random sampling and bias in sampling.

    Well-known polls such as Gallup, Nielsen, Current Population Survey.

    Confidence intervals for percentages in polls.

Introduction to hypothesis testing (very informal).

Rare events.

One-way chi-square test.

Setting up experiments.

References (see also I.M4)

G.M. Clarke and D. Cooke, *A Basic Course in Statistics*, Edward Arnold, London, 1978.

F.N. David, *Games, Gods, and Gambling*, Griffin, London, 1969.

S.E. Hodge and M.L. Seed, *Statistics and Probability*, Blackie & Chambers, London, 1978.

Alan Hoffer, *Statistics and Information Organization*, Creative Publication, Menlo Park, CA.

Myles Holland and Frank Proschan, *The Statistical Exorcist: Dispelling Statistical Anxiety*, Marcel Dekker, New York, 1984.

Robert Hooke, *How to Tell the Liars from the Statisticians*, Marcel Dekker, New York , 1983.

Darrell Huff, *How to Lie with Statistics*, Norton, New York, 1954.

Darrell Huff, *How to Take a Chance,* Penguin, Harmondsworth, UK, 1970.

D.S. Moore, *Statistics: Concepts and Controversies*, W.H. Freeman, San Francisco, 1979.

M.J. Moroney, *Facts from Figures*, Penguin, Harmondsworth, UK, 1951.

Organizing Data and Dealing with Uncertainty, NCTM, Reston, VA, 1979.

W.J. Reichmann, *The Use and Abuse of Statistics*, Penguin, Harmondsworth, UK, 1970.

Report on the Teaching of Probability and Statistics, Mathematics Association, Leicester, UK, 1975.

Terry Sincich, *Statistics by Example*, Dellen Publishing, San Francisco and Santa Clara, CA, 1982.

P. Sprent, *Statistics in Action*, Penguin, Harmondsworth, UK, 1977.

Judith M. Tanur, Frederick Mosteller, William H. Kruskal, Richard F. Link, Richard S. Pieters, Gerald R. Rising and E.L. Lehmann, *Statistics: A Guide to the Unknown*, 2nd edition, Wadsworth and Brooks-Cole, Monterey, CA, 1978.

Kenneth Travers, *et al, Using Statistics*, Addison-Wesley, Reading, MA.

II.M5 Concepts of Algebra in the Middle School

A previous college algebra course is assumed: for example,

Course 4: Algebra and Computing for Elementary and Middle School
Teaching

on pages 19 through 23 of the 1983 RECOMMENDATIONS.

The focus is on algebra as a study of patterns, with the use of variables to express those patterns, and as a study of relationships expressed by equations and inequalities. An historical perspective including the development of algebraic symbolism and methods of solving equations. The change from the view of algebra as generalized arithmetic and the art of solving equations to the view of algebra as the study of mathematical structures is stressed.

Topics especially appropriate for middle school teachers include:

Role and use of variables.

Symbols and syntax.

Ratio and proportion.

Building equations to represent quantitative relations, mappings, and functions (linear, growth, step functions).

Graphical representations.

Equivalences and identities.

Recognition and expression of patterns, including table building and guess and check strategies.

Different ways of representing the same information (equations, graphs, tables).

The relationship of the algebraic concepts in grades 5-9 to those in later grades.

References (see also **I.M5**)

Arthur Hallenberg (ed.), *Historical Topics in Algebra*, NCTM, Reston, VA, 1971.

II.M6 Discrete Mathematics

An introduction to topics in discrete mathematics, with an emphasis on combinatorial problem solving used in a variety of applications.

A *few* of the many following topics, each explored in some depth, may best serve to develop an appreciation of the types of problems and methods of solution encountered in discrete mathematics.

Sets and one-to-one correspondence.

Countability and uncountability.

Inductive proofs and definitions.

Proof by contradiction.

Statements, connectives, and symbolic language.

Truth tables, tautologies, and contradictions.

Boolean functions, logic circuits, Karnaugh maps.

Equivalence relations.

Partially ordered sets.

Pigeonhole principle.

Permutations, combinations.

Binomial and multinomial coefficients.

Recurrence relations, generating functions.

Principle of inclusion-exclusion.

Graphs, connectedness, Euler and Hamilton paths, isomorphism of graphs, trees, planar graphs.

Binary searches, Huffman codes.

Networks, minimum spanning trees, shortest path.

Finite state machines.

References

Bela Bollobas, *Graph Theory: An Introductory Course*, Graduate Texts in Mathematics 63, Springer-Verlag, New York, 1979.

Richard A. Brualdi, *Introductory Combinatorics*, North-Holland, New York, 1977.

John A. Dossey, Alberto D. Otto, Lawrence E. Spence, and Charles Vanden Eynden, *Discrete Mathematics*, Scott Foresman, Glenview, IL, 1987.

Harold Dorwart, Configurations: a case study in mathematical beauty, *Mathematics Intelligencer*, 7 (1985) 39-48.

John N. Fujii, *Puzzles and Graphs*, NCTM, Reston, VA, 1966.

Israel Grossman and Wilhelm Magnus, *Groups and Their Graphs*, New Mathematics Library 14, MAA, Washington DC, 1964.

Franz E. Hohn, *Applied Boolean Algebra: An Elementary Introduction*, 2nd edition, Macmillan, New York, 1966.

Ivan Niven, *Mathematics of Choice*, New Mathematical Library 15, MAA, Washington DC, 1965.

Oystein Ore, *Graphs and Their Uses*, New Mathematical Library 10, MAA, Washington DC, 1963.

G. Ringel, *Map Color Theorem*, Springer-Verlag, New York, 1974.

Steven Roman, *An Introduction to Discrete Mathematics*, Saunders College Publishing, East Sussex, UK, 1986.

Herbert John Ryser, *Combinatorial Mathematics*, Carus Mathematics Monographs 14, MAA, Washington DC, 1963.

Robin J. Wilson, *Introduction to Graph Theory*, Oliver & Boyd, Edinburgh, 1972.

DESCRIPTIONS FOR MATHEMATICS EDUCATION COURSES FOR LEVEL II
(Read also the introductory statement "Mathematics Education Courses for Level III.")

II.ME1 Teaching Strategies in Middle School Mathematics

The general principles are outlined in some detail in the course description for **I.ME1**. At Level II, however, the primary focus must be on the transition from elementary to secondary school mathematics. Attention must be given to the way the adolescent learner interacts with problems of motivation and to finding an appropriate balance between concept learning and skill building.

In addition to the basic topics, the Level II teacher needs a variety of enrichment materials. Puzzle and problem sections in national publications and a steady flow of recreational mathematics books provide ample sources. A few examples of topics are given here; many more will suggest themselves.

Construction with ruler and compasses.

Paper folding and the making of polyhedra.

Positional games.

Coordinate geometry via the game of Battleships.

Analysis of Tic-Tac-Toe, Nim, and Chomp.

Puzzles, paradoxes, and mathematical magic tricks.

Fibonacci numbers.

Tesselations.

Map coloring.

Logic problems.

Use of software.

References (see also **I.ME1** and **II.M1**)

D. Bell, E.R. Hughes and J. Rogers, *Area, Weight and Volume: Monitoring and Encouraging Children's Conceptual Development*, Nelson, London, 1975.

Robert B. Davis, *Discovery in Mathematics: A Text for Teachers*, Cuisenaire Company of America, New Rochelle, NY, 1964.

Robert B. Davis, *Exploration in Mathematics: A Text for Teachers*, Cuisenaire Company of America, New Rochelle, NY, 1967.

Kenneth E. Easterday, Loren L. Henry, and F. Morgan Simpson (eds.), *Activities for Junior High School and Middle School Mathematics: Readings from The Arithmetic Teacher and The Mathematics Teacher*, NCTM, Reston, VA, 1981.

Gerald H. Elgasten and Albert S. Posamentier, *Using Computers in Programming and Problem Solving*, Addison-Wesley, Menlo Park, CA, 1984.

Martin Gardner, *Sixth Book of Mathematical Games from Scientific American*, Charles Scribner's Sons, New York, 1971.

Martin Gardner, *The Unexpected Hanging and Other Mathematical Diversions*, Simon and Schuster, New York, 1969.

Martin Gardner, *Mathematical Carnival*, Alfred A. Knopf, New York, 1975.

Martin Gardner, *Mathematical Puzzles of Sam Loyd*, Dover, New York, 1959.

Alan Hoffer (ed.), *Number Sense and Arithmetic Skills; Ratio, Proportion, and Scaling; Geometry and Visualization; Mathematics in Science and Society; Statistics and Information Organization; Didactics and Mathematics: The Art and Science of Learning and Teaching Mathematics*, Creative Publications, Palo Alto, CA 1978.

George Immerzeel and Bob Wills (eds.), *Ideas from The Arithmetic Teacher; Grades 4-6 Intermediate*, NCTM, Reston, VA, 1979.

George Immerzeel and Melvin L. Thomas (eds.), *Ideas from The Arithmetic Teacher; Grades 6-8 Middle School*, NCTM, Reston, VA, 1982.

Robert J. Kelley, Chomp--An introduction to definitions, conjectures and theorems, *The Mathematics Teacher*, 79 (1986) 516-519.

Bonnie Litwiler and David Duncan, *Activities for the Maintenance of Computational Skills and the Discovery of Patterns*, NCTM, Reston, VA, 1980.

Merle Mitchell, *Mathematical History: Activities, Puzzles, Stories and Games*, NCTM, Reston, VA, 1978.

Janet Morris, *How to Develop Problem Solving Using a Calculator*, NCTM, Reston, VA, 1981.

M. Preston (ed.), *Why, What, and How? Some Basic Questions for Mathematics Teaching*, The Mathematics Association, Leicester, UK, 1976.

Sidney Sharon and Robert E. Reys (eds.), *Applications in School Mathematics*, 1979 Yearbook, NCTM, Reston, VA.

Linda Silvey and James R. Smart (eds.), *Mathematics for the Middle Grades (5-9)*, 1982 Yearbook, NCTM, Reston, VA.

M.N. Suydam and Robert E. Reys (eds.), *Developing Computational Skills*, 1978 Yearbook, NCTM, Reston, VA.

K.J. Travers, L. Pikart, M.N. Suydam and G.E. Runion, *Mathematics Teaching*, Harper and Row, New York, 1977.

F.R. Watson, *Developments in Mathematics Teaching*, Open Books, Somerset, UK, 1976.

II.ME2 Current Issues and Research in Middle School Mathematics

Compare the description of **I.ME2**. Research on learning theories is applied to the teaching of mathematics in grades 5-9.

Relation of curriculum to entire K-12 curriculum.

Changing emphases in curriculum.

The impact of technology.

Computer assisted instruction.

Classroom structure and management.

Grouping for instruction.

Use of a mathematics laboratory.

Individualized instruction.

Provision for student differences.

Enrichment versus acceleration.

Techniques of evaluation.

References

Mark Driscoll, *Stories of Excellence: Ten Case Studies from a Study of Exemplary Mathematics Programs*, NCTM, Reston, VA, 1987.

Curtis McKnight, *et al*, *The Underachieving Curriculum: Assessing School Mathematics from an International Point of View*, Stipes, Champaign, IL, 1987.

Jack Price and J.D. Gawronski (eds.), *Changing School Mathematics: A Responsive Process*, NCTM, Reston, VA, 1981.

Harold L. Schoen and Marilyn J. Zweng (eds.), *Estimation and Mental Computation*, 1986 Yearbook, NCTM, Reston, VA.

Richard J. Shumway, *Research in Mathematics Education*, NCTM, Reston, VA, 1980.

Edward A. Silver (ed.), *Teaching and Learning Mathematical Problem Solving: Multiple Research Perspective*, Lawrence Erlbaum, Hillsdale, NJ, 1987.

Lynn Arthur Steen, Mathematics Education: A Predictor of Scientific Competitiveness, *Science*, 237 (17 July, 1987), 251-2, 302.

II.ME3 Diagnosis and Remediation in Middle School Mathematics

Compare the description for **I.ME3**. At Level II, relevant topics for experiments and activities for the diagnosis and remediation of student difficulties include:

Language related errors.

Fractions and decimals.

Measurement and estimation.

Variables and equations.

Graphing.

Informal Geometry.

Spatial relations.

Probability and statistics.

References (see also I.ME3)

W.C. Adamson, *Handbook for Specific Learning Difficulties*, Gardner's Press, East Sussex, UK, 1979.

W.J. Bush and K.W. Waugh, *Diagnosing Learning Difficulties*, Merrill, Columbus, OH, 1976.

Jon M. Englehardt, Math Clinics, in *Diagnostic and Prescriptive Mathematics: Issues, Ideas, and Insights*, 1984 Research Monograph, Research Council for Diagnostic and Prescriptive Mathematics, Stillwater, OK.

J. Francis-Williams, *Children with Specific Learning Difficulties*, Pergamon, Oxford, UK, 1974.

William C. Lowry (ed.), *The Slow Learner in Mathematics*, NCTM, Reston, VA, 1972.

C. Gains and J. McNicholas, *Remedial Education--Guidelines for the Future*, Longmans, White Plains, NY, 1979.

F.K. Reisman and S.H. Kaufman, *Teaching Mathematics to Children with Special Needs*, Merrill, Columbus, OH, 1980.

II.ME4 The Psychology of Learning Mathematics

See the description of **I.ME4**. The following references may be particularly appropriate for Level II.

References (see also **I.ME4**)

George Bright, John Harvey and Margaret Wheeler, *Learning and Mathematics Games*, Journal for Research in Mathematics Education Monograph, NCTM, Reston VA, 1985.

Robert Case, Implications of cognitive science for instruction, in T.A. Romberg (chair), *School Mathematics: Options for the 1990's*, Proceedings of a Conference at the U.S. Office of Education, Washington DC, 1984.

F. Joe Crosswhite, Jon L. Higgins, Alan Osborne and Richard Shumway, *Teaching Mathematics: Psychological Foundations*, Wadsworth, Belmont, CA, 1973.

Vincent J. Glennon (ed.), *The Mathematical Education of Exceptional Children and Youth: An Interdisciplinary Approach*, NCTM, Reston, VA, 1981.

Jeremy Kilpatrick, Izak Wirszup, Edward Begle and James Wilson (eds.), *Soviet Studies in the Psychology of Learning and Teaching Mathematics*, 14 volumes, NCTM, Reston, VA, 1969-1975.

Curtis C. McKnight, *et al*, *The Underachieving Curriculum: Assessing U.S. School Mathematics from An International Perspective*, Stipes, Champaign, IL, 1987.

Marvin Minsky, *The Society of the Mind*, Simon & Schuster, New York, 1986.

Richard R. Skemp, *The Psychology of Learning Mathematics*, 2nd edition, Penguin, Harmondsworth, UK, 1986.

Lynn Arthur Steen, Mathematics education: A predictor of scientific competitiveness, *Science*, 237 (17 July, 1987) 251, 252, 302.

P.M. van Hiele, *Structure and Insight: A Theory of Mathematics Education*, Academic Press, New York, 1986.

MASTER'S PROGRAMS FOR TEACHERS OF HIGH SCHOOL MATHEMATICS (LEVEL III)

These programs are designed to

• Give teachers renewed enthusiasm for teaching mathematics.

• Provide for professional growth and development.

• Improve teachers' skill in problem solving and in the teaching of problem solving.

• Refresh, update, and extend their knowledge of mathematics.

Before embarking on such a program, participants need to have a sound undergraduate preparation for teaching secondary school mathematics, such as the Level III program described in the 1983 RECOMMENDATIONS.

The program consists of at least 30 semester hours, including at least 24 semester hours in mathematics and at least 6 semester hours in mathematics education. The overall breadth and quality of the program are of critical importance. Selection of courses will vary, but will normally include at least one course in each of the following areas, culminating in a seminar such as **III.M16.**

• Geometry: combinatorial, differential, projective, non-Euclidean, or transformational, and its historical development; e.g., **III.M2, III.M3, III.M4.**

• Continuous mathematics: calculus, analysis, topology, and their historical development; e.g., **III.M12, III,M13.**

• Discrete mathematics: graph theory, applied group theory, generating functions, combinatorics, matrix theory and applications; e.g., **III.M5, III.M6, III.M8, III.M9, III.M10, III.M11.**

• Computing: numerical analysis, combinatorics, or a course (statistics, probability, linear algebra, number theory, group theory) that makes extensive use of the computer as a tool; e.g., **III.M7.**

• Computer Science, as opposed to computing: the design, development, and testing of programs for use in teaching secondary school mathematics; e.g., **III.ME5.**

• Problem solving, statistics, and mathematical modelling; e.g., **III.M1, III.M14, III.M15.**

• Strategies for teaching mathematics; e.g., **III.ME1, III.ME3.**

• Mathematics education, psychology of learning mathematics, the mathematics curriculum, design of courses for students with special needs; e.g., **III.ME2, II.ME4.**

DESCRIPTIONS FOR MATHEMATICS COURSES FOR LEVEL III

Many of the topics for these sample courses have been chosen because they are *not* normally covered in a bachelor's program in mathematics. Although preparation for advanced study is not the main intent, the approach is not less rigorous. The emphasis is on understanding, on foundations, and on exposition rather than an immediate rush to the frontiers of research. The statement of intelligible unsolved problems breathes life into any course. The choice of, and approach to, topics in the courses and the methods used in teaching these courses are dictated by the intent of the program. Outstanding instructors who are willing to develop especially appropriate courses and who enjoy teaching must be selected to teach in this master's program. Teachers are particularly interested in courses that are intellectually stimulating, challenging, and tailored to their professional interests. Regardless of the strength of the teachers' mathematical background it is important to recognize that since college graduation, which may have been many years ago, they may have dealt only with mathematics at or below the precalculus level. Thus both their mathematics and their study skills in mathematics can be enhanced by sympathetic and supportive teaching.

To serve the needs of teachers, careful attention must be given to the manner in which the courses are taught. Each course includes considerable opportunity for class discussion. Since participants are usually experienced teachers, they often should present some topics to the class. This approach encourages discussion since teachers are more willing to listen to, and be critical of, their peers. There is a significant amount of homework which is carefully appraised and criticized. Extensive use of examples, applications and models, introduced in a problem-solving approach, help to encourage explorations and inductive thinking. Active participation helps the teachers develop a better appreciation for doing mathematics and provides examples of teaching techniques which the teachers can later emulate as they teach their own classes. Each of the following courses is offered in this spirit.

The format of the course descriptions varies. Where the amount of the material is likely to be more than enough for a course, it has been partitioned into boldfaced **Topics**. Select a few **Topics** and pursue them in fair depth, rather than give a cursory and superficial overview of the whole subject.

III.M1 Problem Solving in Secondary School Mathematics

See the descriptions of **I.M1**, and **II.M1**. In **III.M1**, problems are selected from number theory, geometry, analytic geometry, probability, statistics, algebra, discrete mathematics, logic, and calculus.

Strategies for problem solving.

Asking the right questions.

Stating the problem in a different way.

Special cases and pattern recognition.

Generalization.

Specialization and analogy.

Working backwards.

Using plausibility arguments.

Counterexamples and proofs.

References (see also **I.M1** and **II.M1**)

Thomas Butts, *Problem Solving in Mathematics*, Scott Foresman, Glenview, IL, 1972.

A. Gardiner, *Discovering Mathematics: The Art of Investigation*, Oxford University Press, London, 1987.

Martin Gardner, *Aha! Gotcha*, W.H. Freeman, San Francisco, 1977.

Martin Gardner, *Aha! Insight*, W.H. Freeman, San Francisco, 1978.

Stephen Krulik and Jesse A. Rudnick, *A Source Book for Teaching Problem Solving*, Allyn and Bacon, Boston, 1984.

Stephen Krulik and Jesse A. Rudnick, *Problem Solving; A Handbook for Teachers*, Allyn and Bacon, Boston, 1980.

George Pólya, *Mathematics and Plausible Reasoning*, Vols. 1 and 2, Princeton University Press, Princeton, NJ, 1954.

George Pólya, *Mathematical Discovery*, Vols. 1 and 2, John Wiley and Sons, New York, 1962.

Oscar Schaaf, *Lane County Mathematics Project: Problem Solving in Mathematics*, Dale Seymour, Palo Alto, CA, 1983.

Alan H. Schoenfeld, *Problem Solving in the Mathematics Curriculum*, MAA Notes Number 1, MAA, Washington DC, 1983 (contains an extensive bibliography).

Wayne A. Wickelgren, *How to Solve Problems*, W.H. Freeman, San Francisco, 1974.

III.M2 Geometry: An Historical Approach

The uses of geometry in astronomy, telling time, navigation, surveying, architecture, art; other current uses.

The historical development of geometric ideas, with a comparison of synthetic (axiomatic), analytic (coordinate), and vector approaches for figures in two and three dimensions.

The use of mathematical models and intuitive discussion prior to rigorous arguement in the development of geometry, with brief consideration of representations involving geoboards, paper folding, and classical constructions.

The development of geometry as a mathematical discipline: axiomatics, recognition of other geometries (the non-Euclidean geometries, spherical geometry, topology), the Erlanger Program (projective geometry, affine geometry, the geometries of similar figures and of equiareal figures, Euclidean geometry), a transformational approach to Euclidean geometry, and symmetry.

Applications of computer graphics to the study of geometry, especially in three or more dimensions.

References (see also III.M3 and III.M4)

L.M. Blumenthal, *A Modern View of Geometry*, W.H. Freeman, San Francisco, 1961.

J.L. Coolidge, *A History of Geometrical Methods*, Dover, New York, 1963.

O.A.W. Dilke, *Mathematics and Measurement, Reading the Past*, Vol. 2, University of California, British Museum, London, 1987.

Howard Eves, *A Survey of Geometry*, Vols. 1 and 2, William C. Brown, Dubuque, IA, 1963.

T.J. Fletcher (ed.), *Some Lessons in Mathematics*, Cambridge University Press, London.

K.O. Friedrichs, *From Pythagoras to Einstein*, New Mathematical Library 16, MAA, Washington DC, 1965.

David Gans, *Transformations and Geometries*, Appleton-Centry-Crofts, New York, 1969.

Ellery B. Golos, *Foundations of Euclidean and non-Euclidean Geometry*, Holt, Rinehart, and Winston, New York, 1968.

Heinrich W. Guggenheimer, *Plane Geometry and Its Groups*, Holden-Day, San Francisco, 1967.

Sir Thomas L. Heath, *A Manual of Greek Mathematics*, Dover, New York.

Michael J. Holt, *Mathematics in Art*, Van Nostrand, Princeton, NJ, 1971.

M. Jeger, *Transformation Geometry*, George Allen and Unwin, London, 1966.

Edwin Moise, *Elementary Geometry from an Advanced Standpoint*, Addison-Wesley, Reading, MA, 1962.

Otto Neugebauer, *The Exact Sciences in Antiquity*, 2nd edition, Brown University Press, Providence, RI, 1970.

Alfred S. Posamentier, *Excursions in Advanced Euclidean Geometry*, rev. edition, Addison-Wesley, Menlo Park, CA, 1984.

Walter Prenowitz and Meyer Jordan, *Basic Concepts of Geometry*, Blaisdell, New York, 1965.

G.Y. Rainich and S.M. Dowdy, *Geometry for Teachers*, John Wiley and Sons, New York, 1968.

Constance Reid, *A Long Way From Euclid*, Routledge and Kegan, New York, 1973.

Garth E. Runion and James R. Lockwood, *Deductive Systems: Finite and non-Euclidean Geometries*, NCTM, Reston, VA, 1978.

I.M. Stewart, *Concepts of Modern Mathematics*, Penguin, Harmondsworth, UK, 1975.

Patrick J. Tyan, *Euclidean and Non-Euclidean Geometry*, Cambridge University Press, London, 1986.

I.M. Yaglom, *Geometric Transformations, I, II, III* (transl. Allen Shields, Abe Shenitzer) New Mathematical Library 8, 21, 24, MAA, Washington DC, 1962, 1968, 1973.

III.M3 Geometry: A Polyhedral Approach

Construction and study of polygons, polyhedra and (two and three-dimensional projections of) polytopes and their symmetry groups.

Star-polygons and Kepler-Poinsot polyhedra.

Regular and semi-regular tilings of the plane. Archimedean polyhedra.

Euler's formula. Dehn-Somerville equations.

The formula of Descartes for the sum of the angular deficiencies on a polyhedron; Pólya's proof of its relations with Euler's formula.

Convexity. Duality. Rigidity. Deformable polyhedra.

Hexaflexagons, rotating rings of tetrahedra, collapsoids.

Geometry in architecture, art, biology, chemistry, and crystallography.

References (see also **III.M2** and **III.M4**)

Edwin A. Abbott, *Flatland*, 1885, 6th edition, Dover, New York, 1952.

Friedrich Bachmann and Eckart Schmidt, *n-gons* (transl. Cyril W. L. Garner), University of Toronto Mathematical Expositions, Number 18, Toronto, Canada, 1975.

Robert Connelly, Flexing Surfaces, in David A. Klarner (ed.), *The Mathematical Gardner*, Prindle, Weber, and Schmidt, Boston, 1981.

H.S.M. Coxeter, *Twisted Honeycombs*, CBMS Number 4, American Mathematical Society, Providence, RI, 1970.

H.S.M. Coxeter, Angels and Devils, in D.A. Klarner (ed.), *The Mathematical Gardner*, Prindle, Weber, and Schmidt, Boston, 1981.

H. Martyn Cundy and A.P. Rollett, *Mathematical Models*, Clarendon Press, Oxford, UK, 1951.

A. Ehrenfeucht, *The Cube Made Interesting*, Pergamon Press, Oxford, UK, 1964.

Adrien L. Hess, *Four Dimensional Geometry - An Introduction*, NCTM, Reston, VA, 1977.

David Hilbert and S. Cohn-Vossen, *Geometry and the Imagination*, Chelsea, New York, 1952.

Peter Hilton and Jean Pedersen, *Build Your Own Polyhedra*, Addison-Wesley, Menlo Park, CA, 1988.

Nicholas D. Kazarinoff, *Geometric Inequalities*, New Mathematical Library 4, MAA, Washington DC, 1952.

I. Lakatos, *Proofs and Refutations: The Logic of Mathematical Discovery*, Cambridge University Press, London, 1976.

Caroline H. MacGillavry, *Symmetry Aspects of Escher's Periodic Drawings*, 2nd edition, Scheltman and Hokema, Bohn, 1976.

Phares G. O'Daffer and Stanley R. Clemens, *Geometry: An Investigative Approach*, Addison-Wellsley, Reading, MA, 1977.

Doris W. Schattschneider and W. Walker, *M. C. Escher Kaleidocycles*, Ballantine, New York, 1977.

Herbert Taylor, Bicycle tubes inside out, in D.A. Klarner (ed.), *The Mathematical Gardner*, Prindle, Weber and Schmidt, Boston, 1981.

L. Fejes Tóth, *Regular Figures*, Pergamon Press, Oxford, UK, 1964.

Magnus J. Wenninger, *Polydron Models*, NCTM, Reston, VA, 1975.

Hermann Weyl, *Symmetry*, Princeton University Press, Princeton, NJ, 1952.

III.M4   Geometry: A Combinatorial and Topological Approach

Euler's formula and applications.

Map coloring and networks. The four and five color theorems.

Jordan curve theorem. Möbius strips. Klein bottles.

Topological equivalence of surfaces. Classification of surfaces.

Map coloring on surfaces of higher genus. The Ringel-Youngs theorem.

Knots.

The Redfield-Pólya-de Bruijn enumeration theorem.

References (see also **III.M2** and **III.M3**)

S. Barr, *Experiments in Topology*, John Murray, London, 1965.

Norman L. Biggs, E. Keith Lloyd and Robin J. Wilson, *Graph Theory, 1736-1936*, 2nd edition, Oxford University Press, London, 1987.

W.G. Chinn and Norman E. Steenrod, *First Concepts of Topology*, New Mathematical Library 18, MAA, Washington DC, 1966

H.S.M. Coxeter, *Introduction to Geometry*, 2nd edition, John Wiley and Sons, New York, 1969.

H.S.M. Coxeter and S.L. Greitzer, *Geometry Revisited*, New Mathematical Library 19, MAA, Washington DC, 1967.

André Delachet, *Contemporary Geometry*, Dover, New York, 1962.

M.C. Escher, *The Graphic Work of M.C. Escher*, Oldbourne Press, London, 1961.

K.J. Falconer, *The Geometry of Fractal Sets*, Cambridge University Press, London, 1985.

Branko Grünbaum and G. C. Shepard, *Tilings and Patterns*, W.H. Freeman, New York, 1987.

Nicholas D. Kazarinoff, *Geometric Inequalities*, New Mathematical Library 4, MAA, Washington DC, 1961.

D.A. Klarner (ed.), *The Mathematical Gardner*, C.J. Bouwkamp, Packing handled pentacubes, pp. 234-242; Stefan A. Burr, Planting trees, pp. 90-99; Branko Grunbaum and G. C. Shephard, Some problems on plane tiling, pp. 167-196; Dean G. Hoffman, Packing problems and inequalities, pp. 212-225; David A. Klarner, Any life among the polyominoes, pp. 243-262; Raphael M. Robinson, Can cubes avoid meeting face to face? pp. 225-233; Doris W. Schattschneider, In praise of amateurs, pp. 140-166; Prindel, Weber and Schmidt, Boston, 1982.

H. Lindgren and G. Fredrikson, *Geometric Disections*, Dover/Constable, New York, 1965.

III.M5 Theory of Functions and Equations

Technology is changing the approach to classical topics in theory of equations. Symbol manipulator and graphing utility software as well as other developing software and calculator techniques increase the importance of the topics listed below while restructuring the emphases in content and teaching techniques.

The algebra of polynomials.

Algebraic equations and their roots; functions and their zeros.

The remainder and factor theorems.

Synthetic division.

The fundamental theorem of algebra.

Relationship between roots and coefficients.

Expansion of $xf'(x)/f(x)$ giving sums of powers of roots.   (Burnside and Panton, Vol. 1, problem 2, p. 172.

Rational roots.

Real and complex roots.

Location of roots.

The Descartes rule of signs.

Cubic and biquadratic equations. Cardano. Tartaglia.

Approximate values of roots. Newton's method. Error terms.

Discriminants, the Vandermonde determinant.

Many of the printed references listed below are out-of-print. None incorporate current approaches to the topics that take advantage of currently available technology. There is a profound need for new printed materials that show the relationship between new technology-based techniques and these classical mathematical topics that play such an important role in secondary school mathematics.

References

William Snow Burnside, Arthur William Panton, *Theory of Equations*, 7th edition, Vols. 1 and 2, Dublin University Press, Dublin, Ireland, 1912.

B.F. Caviness, Computer algebra: Past and future, *Journal of Symbolic Computation*, 2 (1986),   217-236.

B.W. Char, K.O. Geddes, and G.H. Gonnet, *Maple User's Manual*, University of Waterloo Research Report CS-83-41, December, 1983.

Leonard Eugene Dickson, *Elementary Theory of Equations*, John Wiley and Sons, London, 1914.

John M. Hosack, A guide to computer algebra systems, *The College Mathematics Journal*, (17 November, 1986) 434-441.

MACSYMA, Symbolics Inc., Cambridge, MA, 1988.

Bruce E. Meserve, *Fundamental Concepts of Algebra*, Dover, New York, 1981.

The muMATH/muSIMP-80 Symbolic Mathematics System Reference Manual for the Apple II Computer, The Soft Warehouse, Honolulu, 1981.

Yves Nievergelt, The chip with the college education: the HP-28C, *American Mathematical Monthly*, 94 (November, 1987) 895-902.

Joseph M. Thomas, *Theory of Equations*, McGraw-Hill, New York, 1938.

I. Todhunter, *An Elementary Treatise on the Theory of Equations*, 4th edition, Macmillan, New York, 1985.

H.W. Turnbull, *Theory of Equations*, Interscience, New York, 1939.

J.V. Uspensky, *Theory of Equations,* McGraw-Hill, New York, 1948.

Bert K. Waits and Franklin Demana, Editorial: Solving problems graphically using microcomputers, *UMAP Journal* , 8 (1987) 1-7.

Herbert S. Wilf, The disk with the college education, *American Mathematical Monthly*, 89 (January, 1982) 4-8.

III.M6 Topics in Number Theory

This is intended as a second course, having the same prerequisites and objectives as, but otherwise independent of the course described on pages 54 through 56 of the 1983 RECOMMENDATIONS. There is a wide variety of possible topics. Five examples are given, but only two, or at most three, should be selected, and then studied in fair detail. The elementary proof of the prime number theorem is an accessible topic, but hard to make exciting. See, however, the Levinson reference below.

Partitions A topic that uses tools from combinatorics (Ferrers, Sylvester, Durfee, MacMahon), algebra (Euler and the formal manipulation of series and products), and analysis (Jacobi, theta functions, Hardy-Ramanujan-Rademacher). Proofs of a few theorems by several different methods.

Ferrers's diagrams. Conjugate partitions. Durfee square.

Generating functions for unrestricted and restricted partitions.

Euler's pentagonal number identity. Rogers-Ramanujan identities.

Relationship of partitions with the sum-of-divisors function.

Jacobi's identity. Theta functions can be treated with a minimum of complex variables (state theorem on residues and Liouville's theorem).

The thrilling story of the Hardy-Ramanujan asymptotic formula (see Littlewood references below). Rademacher's improvement.

Continued Fractions   Finite and infinite simple continued fractions. Partial quotients. Convergents. Complete quotients. Linear diophantine equations in two unknowns. Approximations of irrational numbers by rational numbers. Hurwitz's theorem. Equivalence of quadratic surds and their expansions as periodic continued fractions. Applications to Pell equations.

Algebraic Number Fields   Algebraic and transcendental numbers. Liouville's construction. Algebraic integers. Units. Quadratic fields and their rings of integers. Norms. Conjugates. Primes. Euclidean fields. Unique factorization. Gaussian and Eisenstein fields. Proof of Fermat's last theorem when $n = 3$.

Binary Quadratic Forms   Discriminant. Equivalence. Both proper and primitive representations. Reduction of positive definite forms. Number of representations. Composition of forms. Class number.

Recurring Sequences   Second order recurrences. Fibonacci, Lucas, and Pell sequences as special cases. Lucas-Lehmer law of repetition. Law and rank of apparition. Divisibility properties. Primitive and intrinsic factors. Lucas-Lehmer primality test for Mersenne numbers. Periodic and degenerate sequences.

References

Alan Baker, *A Concise Introduction to the Theory of Numbers*, Cambridge University Press, London, 1984.

A.H. Beiler, *Recreations in the Theory of Numbers: The Queen of Mathematics Entertains*, Dover, New York, 1966.

H. Davenport, *The Higher Arithmetic*, 5th edition, Cambridge University Press, London, 1982, (Ch. 4 Continued Fractions, Ch. 6 Quadratic forms).

G.H. Hardy and S. Ramanujan, *Proceedings of the London Mathematical Society*, 17 (1918) 75-115.

G.H. Hardy and E.M. Wright, *Introduction to the Theory of Numbers*, 4th edition, Oxford University Press, London, 1960.

A. Ya. Khintchin, *Continued Fractions*, P. Noordhoff, Groningen, 1963.

Norman Levinson, A motivated account of the elementary proof of the prime number theorem, *American Mathematical Monthly*, 76 (1969) 225-245.

J.E. Littlewood, *A Mathematician's Miscellany*, Methuen, London, 1953.

Calvin T. Long, *Elementary Introduction to Number Theory*, 3rd edition, Prentice Hall, Englewood Cliffs, NJ, 1987.

Major P.A. MacMahon, *Combinatory Analysis*, Chelsea, 1960 (esp. Vol. 1, pp. 224-287 and Vol. 2, pp. 1-88 for partitions).

Ivan Niven, *Irrational Numbers*, Carus Mathematical Monograph 11, MAA, Washington DC, 1956.

Ivan Niven, *Numbers: Rational and Irrational*, New Mathematical Library 1, MAA, Washington DC, 1961.

Ivan Niven and Herbert S. Zuckerman, *An Introduction to the Theory of Numbers*, 4th edition, John Wiley and Sons, New York, 1980.

C.D. Olds, *Continued Fractions*, New Mathematical Library 9, MAA, Washington DC, 1963.

Oystein Ore, *Invitation to Number Theory*, New Mathematical Library 20, MAA, Washington DC, 1967.

J.J. Sylvester, *Collected Mathematical Papers*, Chelsea, New York, 1973 (Vol. II, p. 86-99, 119-175, 701-703; Vol. III, pp. 661-666, 680-686; Vol. !V, pp. 1-83, 91-100).

N.N. Voroby'ev, *Fibonacci Numbers*, D.C. Heath, Boston, 1963.

III.M7 Numerical Analysis

An algorithmic approach with frequent computer implementation. Constant consciousness of the existence and magnitude of errors, even where formal error analysis is impractical.

Computation The two deadly sins (small differences of large numbers; division by small numbers). Fixed and floating point arithmetic. Rounding and other errors.

Numerical solution of equations Iteration. Convergence. Fixed points. Lipschitz condition. Aitken's method. Newton-Raphson method.

Numerical linear algebra Finite differences. Difference equations and operators. Matrix manipulation. Systems of linear equations. Ill-conditioned matrices. Contraction maps.

Interpolation Polynomial interpolation. Lagrange's formula. Inverse interpolation. Errors. Spline interpolation. Approximation theory. Least squares.

Quadrature Numerical integration. Errors.

Numerical solution of differential equations Numerical differentiation. Ordinary differential equations. Predictors and correctors. Partial differential equations. Boundary value problems.

Alternative topics include:

Finite element methods.

Application to structures.

Control of industrial Operations.

Fast Fourier transforms.

References

BASIC Numerical Methods (An Introduction to Numerical Mathematics on a microcomputer), Arnold, London, 1984.

Richard L. Burden, J. Douglas Faires and Albert C. Reynolds, *Numerical Analysis*, 3rd edition, Prindle, Weber, and Schmidt, Boston, 1985.

J.D. Conte and Carl de Boor, *Elementary Numerical Analysis*, 2nd edition, McGraw-Hill, New York, 1972.

Germund Dehlquist and Åke Björck, *Numerical Methods*, Prentice Hall, Englewood Cliffs, NJ, 1974.

L. Fox, *An Introduction to Numerical Linear Algebra*, Oxford University Press, London, 1964.

Gene H. Golub and Charles F. Van Loan, *Matrix Computations*, Johns Hopkins University Press, Baltimore, MD, 1983.

P. Henrici, *Essentials of Numerical Analysis*, John Wiley and Sons, New York, 1982.

Eugene Isaacson and Herbert Bishop Keller, *Analysis of Numerical Methods*, John Wiley and Sons, New York, 1966.

Anthony Ralson and Philip Rabinowitz, *A First Course in Numerical Analysis*, 2nd edition, McGraw-Hill, New York, 1978.

Terry E. Shoup, *Applied Numerical Methods for the Microcomputer*, Prentice Hall, Englewood Cliffs, NJ, 1984.

G.W. Stewart, *Introduction to Matrix Computations*, Academic Press, New York, 1973.

III.M8 Coding Theory

Channels. Noise. Information theory. Shannon's theorem.

Linear codes. Decoding. Error probability. Hamming codes. Dual (orthogonal) codes.

Non-linear codes. Hadamard and conference matrices. Finite projective geometries.

The Golay code and its connections with the Mathieu group $M24$, the Steiner system $S(5,6,12)$. lattice packing of spheres, and combinatorial games (Möbius, Mogulk, and lexicodes).

Arithmetic modulo an irreducible binary polynomial. Finite fields. BCH codes.

Dual codes and weight distribution.

Codes, designs and perfect codes.

Cyclic codes. Quadratic residue codes.

Gray codes. Chinese rings puzzle. Snake-in-the-box codes.

References

E.R. Berlekamp, *A Survey of Algebraic Coding Theory*, Springer-Verlag, New York, 1972.

I.F. Blake and R.C. Mullin, *The Mathematical Theory of Coding*, Academic Press, New York, 1975.

P.J. Cameron and J.H. van Lint, *Graph Theory, Coding Theory, and Block Designs*, London Mathematical Society Lecture Notes 19, Cambridge University Press, London, 1975.

Solomon W. Golomb (ed.), *Digital Communications with Space Applications*, Prentice Hall, Englewood Cliffs, NJ, 1964.

Shu Lin, *Error Control Coding*, Prentice Hall, Englewood Cliffs, NJ, 1983.

J.H. van Lint, *Coding Theory*, Springer Lecture Notes 201, Springer-Verlag, New York, 1971.

F.J. MacWilliams, and N.J.A. Sloane, *The Theory of Error-Correcting Codes*, North-Holland, New York, 1978 (with a bibliography of over 1500 items).

N.J. Sloane, *A Short Course on Error Correcting Codes*, Springer-Verlag, New York, 1975.

Thomas M. Thompson, *From Error-Correcting Codes Through Sphere Packings to Simple Groups*, Carus Mathematical Monograph 21, MAA, Washington DC, 1984.

III.M9 Combinatorial Games

There is far more accessible material than will go into a single course. Select topics carefully to avoid confusion; five are listed here, with titles of four alternatives.

Impartial Games Nim. Bouton's theory. The Sprague-Grundy theory. Many examples; e.g., Grundy's Game, Kayles, Dawson's Chess. See also Green Hackenbush below. Are all octal games ultimately periodic?

Partisan Games Some games are numbers. Dedekind sections and Cantor's transfinite numbers as special cases. Berlekamp's rule for Blue-Red Hackenbush strings. Infinitesimals. Star, up, down and other "all small" games.

Simplicity Sums (disjunctive compounds) of games. All Games from a partially ordered commutative group. Dominated and reversible options. Simplest forms of games.

Hot Games Switching. Cashing Cheques. Introduction to thermography.

Green Hackenbush The colon principle. The parity principle. The fusion principle. The interplay of nim addition and ordinary addition.

Alternative topics include:

Atomic Weights.

Joins and Unions (conjunctive and selective compounds).

Loopy (infinite) **Games.**

Misere Play (last player losing) of impartial games.

Coin-Turning Games.

References

Lowell W. Beineke and Robin J. Wilson, *Selected Topics in Graph Theory 2*, Academic Press, London, 1983, (Chapter 9, Graphs and Games).

E.R. Berlekamp, J.H. Conway, and Richard K. Guy, *Winning Ways for your Mathematical Plays*, Academic Press, New York, 1982 (mainly Vol. 1, but Vol. 2 contains many additional topics and examples).

Charles L. Bouton, Nim, a game with complete mathematical theory, *Annals of Mathematics*, (2) 3 (1901-1902) 35-39.

John H. Conway, *On Numbers and Games*, Academic Press, London, 1976 (gives alternative Hackenbush treatment from that in Berlekamp, Conway and Guy).

H.S.M. Coxeter and W.W. Rouse-Ball, *Mathematical Recreations and Essays*, 12th edition University of Toronto Press, Toronto, OT, 1974, (pp. 36-40).

P.M. Grundy, Mathematics and games, *Eureka*, 27 (1964) 9-11.

G.H. Hardy and E.M. Wright, *An Introduction to the Theory of Numbers*, 4th edition, Oxford University Press, London, 1960, (pp. 117-120).

III.M10   Graph Theory

There is a wide variety of topics and applications, so select carefully to avoid indigestion. Network flows are omitted since they appear as a separate course, **III.M11**, complementary to this one. Games on a graph (Berge, Chapters 5 and 6) can be studied in course **III.M8**. Five topics are listed here, with titles of four alternatives.

Fundamentals Paths, circuits, chains, cycles. Trees. Connectivity. Hamiltonian cycles. Euler circuits. Menger's theorem.

Matching Tutte's 1-factor theorem. Round-robin tournaments. Matrices of zeros and ones. Application to Sam Loyd's Tantalizer (Instant Insanity).

Planarity Euler's formula. Kuratowski's theorem. Genus, thickness, coarseness, and crossing number.

Coloring Tait's conjecture. The five-color theorem. The Ringel-Youngs theorem. Is the four-color theorem really proved?

Graphs and Groups Cayley and Schreier diagrams. Applications of the adjacency matrix.

Alternative topics include:

Enumeration Burnside's lemma. Redfield-Pólya-de Bruijn theorem.

Extremal Problems (In Chapter 4 of Bollobás, *Graph Theory*).

Ramsey Theory (In Chapter 6 of Bollobás, *Graph Theory*).

Random Graphs (In Chapter 7 of Bollobás, *Graph Theory*).

References (see also **III.M10**)

David Barnette, *Map Coloring, Polyhedra and the Four-Color Problem*, Dolciani Mathematical Expositions 8, MAA, Washington DC, 1984.

Lowell J. Beineke and Robin J. Wilson (eds.), *Selected Topics in Graph Theory*, Vols. 1 and 2, Academic Press, London, 1978 and 1983, (collections of expository articles).

Béla Bollobás, *Extremal Graph Theory*, Academic Press, London, 1978.

Béla Bollobás, *Graph Theory: An Introductory Course*, Springer-Verlag, New York, 1979.

Gary Chartrand and Linda Lesniak, *Graphs and Digraphs*, 2nd edition, Wadsworth and Brooks-Cole, Monterey, CA, 1986.

H.S.M. Coxeter, Roberto Frucht, and David L. Powers, *Zero-Symmetric Graphs*, Academic Press, New York, 1981, (relations between groups and graphs).

D.R. Fulkerson (ed.), *Studies in Graph Theory I and II*, Studies in Mathematics 11 and 12, MAA, Washington DC, 1975 (good link between graph theory and combinatorial optimization).

John J. Fujii, *Puzzles and Graphs*, NCTM, Reston, VA, 1966.

Jack E. Graver and Mark E. Watkins, *Combinatorics with Emphasis on the Theory of Graphs*, Springer-Verlag, New York, 1977.

Israel Grossman and Wilhelm Magnus, *Groups and their Graphs*, New Mathematical Library 14, MAA, Washington DC, 1964.

Frank Harary, *Graph Theory*, Addison-Wesley, Reading, MA, 1969.

Oystein Ore, *Graphs and their Uses*, New Mathematical Library 10, MAA, Washington DC, 1963.

G. Ringel, *Map Color Theorems*, Springer-Verlag, New York, 1974.

Robin J. Wilson, *Introduction to Graph Theory*, Oliver & Boyd, Edinburgh, 1972.

III.M11 Flows in Networks

An exciting field with plenty of applications. Interpret widely, in the sense of combinatorial optimization. A good area to display the **unity** of combinatorics. Computers are invaluable for many topics. Four topics are listed, with titles of four alternatives.

Maximal Flows Sources, sinks, cuts. Disconnecting sets. Multiple sources and sinks. Max-flow min-cut theorem. Integrity theorem.

Combinatorial applications Relation of max-flow min-cut theorem to the König-Egeváry theorem, Menger's graph theorem, Hoffman's unicursal graph theorem, Dilworth's chain decomposition theorem and Hall's marriage theorem.

Linear Programming Duality. Feasibility. Blocking polyhedra.

Task Scheduling A selection from: Hitchcock-Koopmans transportation problem, traveling salesman problem, warehousing problem, caterer problem, knapsack problem, cutting stock problem, assignment problem, bottleneck problem.

Alternative topics include:

Matrices of zeroes and ones.

Systems of Distinct Representatives.

Integer Programming Gomory's method.

Comparison of Algorithms Simplex method. Ellipsoid method. Algorithms of Khachiyan and of Karmarkar.

References (see also III.M9)

G.B. Dantzig, *Linear Programming and Extensions*, Princeton University Press, Princeton, NJ, 1963.

G.B. Dantzig and B.D. Eaves (eds.), *Studies in Optimization*, Studies in Mathematics 10, MAA, Washington DC, 1974.

L.R. Ford and D.R. Fulkerson, *Flows in Networks*, Princeton University Press, Princeton, NJ, 1962.

Alan Gibbons, *Algorithmic Graph Theory*, Cambridge University Press, New York, 1985.

L.G. Khachiyan, A polynomial algorithm in linear programming, *Soviet Mathematics Doklady*, 20 (1979) 191-194.

V.L. Klee and G.L. Minty, How good is the simplex algorithm? in O. Shisha (ed.) *Inequalities III*, Academic Press, New York, 1972.

Eugene L. Lawler, *Combinatorial Optimization Networks and Matroids*, Holt Rinehart and Winston, New York, 1976.

M. O'Leigeartaigh, J.K. Lenstra, and A.H.G. Rinnooy Kan, *Combinatorial Optimization: Annotated Bibliographies*, John Wiley and Sons, New York, 1985.

J. Laurie Snell and Peter Doyle, *Random Walks and Electric Networks*, Carus Mathematical Monograph 22, MAA, Washington DC, 1984.

W.T. Tutte, The quest of the perfect square, *American Mathematical Monthly*, 71 (1976) No. 2, Part II (Slaught Memorial Paper, Number 10) 29-35; (and references therein; application of network flows to a classical problem).

III.M12 History of Calculus

A course that focuses on a limited period in the history of calculus to give depth rather than breadth. In this sample--Archimedes and the period 1650-1750--the ideas and approaches of the original creators are presented in modern notation. The tortuous routes leading to results now accepted as straightforward are traced.

Before Archimedes: Thales, Eudoxus.

Archimedes. Sphere. The "method". Area of parabolic segment. Equilibrium of floating bodies.

Decline during "dark ages". Arabic contributions. Oresme.

Early seventeenth century: Fermat, Descartes, Cavalieri, Kepler, Neil, Barrow, Wallis, Mercator.

Newton.

Leibniz.

Euler (function, power series) and the early eighteenth century.

References

Margaret E. Baron, *The Origins of the Infinitesimal Calculus*, Oxford University Press, London, 1969.

Carl T. Boyer, *A History of the Calculus and its Conceptual Development*, Dover, New York, 1959.

C.H. Edwards, *The Historical Development of the Calculus*, Springer-Verlag, New York. 1982.

Sir Thomas L. Heath (ed.), *The Works of Archimedes*, Dover, New York.

W.M. Priestly, *Calculus: An Historical Approach*, Springer-Verlag, New York, 1979.

W.W. Sawyer, *What is Calculus About?*, New Mathematical Library 2, MAA, Washington DC, 1961.

D.J. Struik, *A Sourcebook in Mathematics, 1200-1800*, M.I.T. Press, Cambridge, MA, 1969.

Otto Toeplitz, *The Calculus, A Genetic Approach*, University of Chicago Press, Chicago, 1963.

D.T. Whiteside (ed.), *The Mathematical Papers of Isaac Newton*, 8 Vols., Cambridge University Press, New York, 1967-81.

D.T. Whiteside, *The Mathematical Works of Isaac Newton*, Vols. 1 and 2, Johnson reprint, New York, 1964.

III.M13 Foundations of Analysis

An in-depth study of central concepts of analysis, with special emphasis on those areas which can enrich or supplement the content of high school courses for advanced mathematics student. This course focuses on the concept of limit as the central theme. Calculators, computers, and graphics are used throughout to explore concepts and make conjectures.

Development of the real numbers and completeness axiom. Dedekind sections, Cauchy sequences or Cantor decimals.

Definition of limit. Role of the completeness axiom in developing this concept. Cauchy, Weierstrass.

Sequences and series. The limit of a sequence, including uniform convergence.

Continuity, including uniform continuity. The relation of the definition of limits as sequences to the definition of continuity.

Definitions of the derivative and the integral.

The Fundamental Theorem of calculus and its proof.

Examples and counterexamples in analysis to distinguish clearly between continuity and differentiability.

Weierstrass's function.

References

Colin Clar, *Elementary Mathematical Analysis*, Wadsworth, Belmont, CA, 1982.

Kenneth A. Ross, *Elementary Analysis: The Theory of Calculus*, Springer-Verlag, New York, 1980.

III.M14 Statistics and Probability

Data analysis.

Data analysis in modern statistics.

Techniques and principles of data analysis including graphical techniques, resistant statistics, transformations.

Probability.

Simulation.

Random number tables and computer random number generators.

Solution of probability problems.

Construction of probability distributions.

Teaching theoretical probability.

Problems people have learning probability (see Tversky).

Successful approaches to conditional probability, to the addition rule, to the multiplication rule, and to permutations and combinations.

The normal distribution.

A model of empirical data.

An approximation to the binomial distribution.

The central limit theorem.

Statistics.

Introduction to sampling surveys and confidence intervals.

Sampling including random sampling, bias in sampling, how sampling can go wrong.

Polls such as Gallup, Nielsen, and Current Population Survey.

Confidence intervals for percentages quoted in polls.

Teaching hypothesis testing.

Testing a proportion.

Testing a mean.

Chi-square tests.

Data collection.

Survey design.

Experimental design including replicability, placebo, effect, control group.

References (see also **I.M4** and **II.M4**)

William F. Cleveland, *The Elements of Graphing Data*, Wadsworth, Monterey, CA, 1985.

David Kahneman, Paul Slovic, and Omos Tzersky, *Judgement Under Uncertainty: Heuristics and Biases*, Cambridge University Press, Cambridge, UK, 1982

David Moore, *An Introduction to the Practice of Statistics*, Freeman, San Francisco, in press. (This book will go with a TV course from COMAP.)

Fredrick Mosteller, *et al, Statistics by Example*, Addison-Wesley, Reading, MA, 1973.

Edward R. Tufte, *The Visual Display of Quantitative Information*, Graphics Press, Cheshire CT, 1983.

III.M15 Mathematical Modelling

Mathematical concepts and techniques from earlier courses are used in modelling real-life problems to bring a new vividness and interest to the ideas.

Numerous problems, often ill-posed, are confronted from various sciences - physical, biological, social, behavioral. Problems are reformulated so that a mathematical approach may be used.

The use of computers in solving problems is encouraged.

Experience is provided with a wide variety of models--deterministic, stochastic, simulation, continuous, discrete, axiomatic.

Possible topics include:

Discrete models from numerical data.	Linear Optimization.
Richardson arms-race model.	Predator-prey analysis.
Epidemiology models.	Markow processes.
Resource management models.	Monte Carlo simulations.
Demographic models.	Job scheduling.
Population growth.	Travelling salesman problem.

References

J.G. Andres and R.R. McLone, *Mathematical Modelling*, Butterworths, London, 1976.

Karl Egil Aubert, Spurious mathematical modelling, *Mathematical Intelligencer*, 6 (1984) 654-60.

A. Battersby, *Mathematics in Management*, Penguin, Harmondsworth, UK, 1966.

M.S. Bell, *Mathematical Uses and Models in our Everyday World*, School Mathematics Study Group, Palo Alto, CA, 1972.

E.A. Bender, *An Introduction to Mathematical Modelling*, John Wiley and Sons, New York, 1978.

D.N. Burghes and A.D. Wood, *Mathematical Models in the Social, Management and Life Sciences*, Ellis Horwood, London, 1980.

D.N. Burghes, *Solving Real Problems with Mathematics*, 2 Vols., Cranfield Press, London, 1981, 1982.

H. Burkhardt, *The Real World and Mathematics*, Blackie, London, 1981.

M. Cross and A.O. Moscardindi, *Learning the Art of Mathematical Modelling*, Ellis Horwood, London, 1985.

Frank Giordano and Maurice Weird, *First Course in Mathematical Modelling*, Wadsworth and Brooks-Cole, Belmont, CA, 1985.

Ramesh Kapadia and Huw Kyffin, *Modelling for Schools/Colleges*, Polytechnic of the South Bank, Research Series, London, 1985.

P. Lancaster, *Mathematics: Models of the Real World*, Prentice Hall, Englewood Cliffs, NJ, 1976.

M.J. Lighthill (ed.), *Newer Uses of Mathematics*, Penguin, Harmondsworth, UK, 1978.

Open University Course Team, *Modelling by Mathematics*, 6 books (18 units), Open University Press, London, 1977.

K. Rektorys, *Survey of Applicable Mathematics*, Iliffe, London, 1969.

F.S. Roberts, *Discrete Mathematical Models with Applications to Social, Biological and Environmental Problems*, Prentice Hall, Englewood Cliffs, NJ, 1976.

T.L. Saaty and J.M. Alexander, *Thinking with Models: Mathematical Models in the Physical, Biological, and Social Sciences*, Pergamon, Oxford, UK, 1981.

III.M16 Seminar: Readings in Mathematics

Participants learn the existence of, if not become familiar with, many of the items in the Matthew P. Gaffney and Lynn Arthur Steen (with the assistance of Paul J. Campbell), *Annotated Bibliography of Expository Writing in the Mathematical Sciences*, MAA, Washington DC, 1976.

Reading is used as a basis for exposition in the seminar. This can be at either of two levels:

The speaker introduces a new area of mathematics or its historical development.

An instructional classroom unit is developed by locating references and identifying points of resonance with the high school curriculum.

By pooling the expositions at the second level, participants collect many items suitable for immediate use in the classroom. Note that:

Writing and exposition are more important than reading.
Reading about mathematics is no substitute for doing it.

References

Irving Adler, *The Impossible in Mathematics*, NCTM, Reston, VA, 1957.

Irving Adler, *Readings in Mathematics*, NCTM, Reston, VA, 1972.

Leon Bowden and M.M. Schiffer, *The Role of Mathematics in Science*, New Mathematical Library 30, MAA, Washington DC, 1984.

Robert L. Brabenec, A required reading program for mathematics majors, *American Mathematical Monthly*, 94 (1987) 366-368 (and see the bibliography therein).

Douglas M. Campbell and John C. Higgins, *Mathematics: People, Problems, Results*, Vols. I-III, Wadsworth International Group, Belmont, CA, 1984.

W.K. Clifford, *The Common Sense of the Exact Sciences*, reprinted Dover, New York.

COSRIMS, *The Mathematical Sciences*, M.I.T. Press, Cambridge, MA, 1969.

R. Courant and H. Robbins, *What is Mathematics?*, Oxford University Press, London, 1941.

Philip J. Davis and Reuben Hersh, *The Mathematical Experience*, Houghton Mifflin, Boston, 1981.

Rod D. Driver, *Why Math?*, Springer-Verlag, New York, 1984.

Galileo Galilei, *Dialogues Concerning Two New Sciences*, Dover, New York.

Jacques Hadamard, *The Psychology of Invention in the Mathematical Field*, Dover, New York, 1954.

G..H. Hardy, *A Mathematician's Apology*, Cambridge University Press, London, 1940.

D.R. Hofstadter, *Gödel, Escher, Bach: An Eternal Golden Braid*, Basic Books, New York, 1979; Penguin, Harmondsworth, UK, 1980.

Ross Honsberger, *Ingenuity in Mathematics*, The New Mathematical Library 23, MAA Washington DC, 1970.

Ross Honsberger, *Mathematical Gems I, II, III*, Dolciani Mathematical Expositions 1, 2, 9, MAA, Washington DC, 1973, 1976, 1985.

Ross Honsberger, *Mathematical Morsels*, Dolciani Mathematical Expositions 4, MAA, Washington DC, 1979.

Leopold Infeld, *Whom the Gods Love: The Story of Evariste Galois*, NCTM, Reston, VA, 1978 (see also the Rothman paper below).

Keith Jones (ed.), *Nine Selected Articles from the New Scientist*, New Science Publications, London.

E. Kasner and J.R. Newman, *Mathematics and the Imagination*, Simon & Schuster, New York, 1940.

Felix Klein, *Elementary Mathematics From an Advanced Standpoint*, Dover, New York, 1939.

M. Kline, *Mathematics in the Modern World*, W.H. Freeman, San Francisco, 1968.

M. Kline, *Mathematical Thought from Ancient to Modern Times*, Oxford University Press, London, 1973.

S. Körner, *Philosophy of Mathematics: An Introductory Essay*, Hutchinson, London, 1971.

J.E. Littlewood, *A Mathematician's Miscellany*, Methuen, London, 1957.

Saunders Mac Lane, *Mathematics: Form and Function*, Springer-Verlag, New York, 1986.

J.R. Newman, *The World of Mathematics*, Vols. I-IV, Simon & Schuster, New York, 1956.

Hans Rademacher and Otto Toeplitz, *The Enjoyment of Mathematics*, Princeton University Press, Cambridge, MA, 1970.

Constance Reid, *Introduction to Higher Mathematics*, Routledge, London, 1960.

Tony Rothman, Genius and biographers: The fictionalization of Evariste Galois, *American Mathematical Monthly*, 89 (1982) 84-106.

Bertrand Russell, *Introduction to Mathematical Philosophy*, George Allen and Unwin, New York, 1919,1960.

William L. Schaaf, *The High School Mathematics Library*, NCTM, Reston, VA, 1982.

Abraham Sinkov, *Elementary Cryptanalysis*, New Mathematical Library 22, MAA, Washington DC, 1966.

Raymond Smullyan, *What is the Name of This Book?*, Penguin, Harmondsworth, UK.

Daniel Solow, *How to Read and Do Proofs*, John Wiley and Sons, New York, 1982.

Lynn Arthur Steen (ed.), *Mathematics Tomorrow*, Springer-Verlag, New York, 1981.

Sherman K. Stein, *Mathematics, The Man-Made Universe*, 3rd edition, W.H. Freeman, San Francisco, 1976.

Hugo Steinhaus, *Mathematical Snapshots*, 3rd edition, Oxford University Press, London, 1942.

MATHEMATICS EDUCATION COURSES FOR LEVEL III

The Mathematics Education course descriptions differ from those for Mathematics courses on one critical point. Many topics in the sample Mathematics courses have been selected because they do not typically appear in undergraduate curricula. For the Mathematics Education courses, many of the topics are listed because they **DO** appear in the initial certification (bachelor's) program. This is because the topic is important enough for teaching that the additional refinement and extension of knowledge about the topic will significantly benefit the teacher. The topic is of sufficient significance that additional research knowledge continues to be generated, and the teacher will encounter new knowledge about teaching and about classroom practices.

It is critical that the mathematics teacher realize that knowledge of how children learn, effective teaching methods, and curricular design, are all constantly changing. Each Mathematics Education course must involve the study of current literature in the field. Journals, including the *American Mathematical Monthly*, *The College Mathematics Journal*, *The Mathematics Magazine*, *The Mathematics Teacher*, *The Arithmetic Teacher*, the *Mathematical Gazette*, the *Journal for Research in Mathematics Education*, *School Science and Mathematics*, and *Educational Studies in Mathematics*, should be readily accessible and used as prime sources in the mathematics education courses. Because the state of the art in methodology, in learning, and in curriculum, is continually changing, the bibliographies for the mathematics education courses are less extensive: the assumption is that the use of current journals in mathematics education is a characteristic of each course.

It is recommended that **III.ME4 The Secondary School Mathematics Curriculum,** be required of all participants, together with at least one other course.

DESCRIPTIONS OF MATHEMATICS EDUCATION COURSES FOR
LEVEL III

III.ME1 A Second Course in Teaching Methods in Secondary School Mathematics

See the descriptions for **I.ME1** and **II.ME1**. An examination of factors affecting teacher effectiveness, including contrasting instructional styles, strategies, and techniques.

Teachers already in service conduct a self-analysis of their own classroom management techniques and teaching skills.

Methodology for specific instructional purposes:

Concept development.   Problem solving.

Skill building.   Proof.

Applications.

Techniques for students in different tracks (general, remedial, college preparatory), with different patterns of reasoning (intuitive, visual, logical, brute force), and with different levels of ability or achievement (capable, gifted, slow learners, underachievers).

Strategies for teaching topics where there may be special difficulties:

Functions.   Rational expressions.

Curve sketching.   Mathematical induction.

Deductive reasoning.   Binomial theorem.

Complex numbers.   Synthetic geometry.

Probability.   Problem solving.

Statistics.

References

Stanley J. Bezuska, Lou d'Angela, and Margaret J. Kenny, *The Wonder Square*, Boston College Mathematics Institute: Motivated Mathematics Project Activity, Booklet 2, Boston College Press, Chestnut Hill, MA, 1976.

Stanley J. Bezuska, Lou d'Angela, and Margaret J. Kenney, *Forever & Ever &...Applications of Series*, Boston College Mathematics Institute: Motivated Mathematics Project Activity, Booklet 4, Boston College Press, Chestnut Hill, MA, 1976.

Stanley J. Bezuska, Lou d'Angela, and Margaret J. Kenney, *Fractive Action*, Boston College Mathematics Institute: Motivated Mathematics Project Activity, Booklet 5, Boston College Press, Chestnut Hill, MA, 1976.

Robert B. Davis, a study in the process of making proofs, *Journal of Mathematical Behavior*, 4 (1985) 37-43.

David R. Johnson, *Every Minute Counts: Making your Math Class Work*, Dale Seymour, Palo Alto, CA, 1982.

David R. Johnson, *Making Minutes Count Even More: A Sequel to Every Minute Counts*, Dale Seymour, Palo Alto, CA, 1986.

Bernice Kastner, *Applications of Secondary School Mathematics*, NCTM, Reston, VA, 1978.

Margaret J. Kenney, *A Lesson in Mathematical Doodling*, Boston College Mathematics Institute: Motivated Mathematics Project Activity, Booklet 12, Boston College Press, Chestnut Hill, MA, 1976.

Margaret J. Kenney, *The Super Sum*, Boston College Mathematics Institute: Motivated Mathematics Project Activity, Booklet 10, Boston College Press, Chestnut Hill, MA, 1976.

Max A. Sobel and Evan M. Maletsky, *Teaching Mathematics: A Sourcebook of Aids, Activities, and Strategies*, 2nd edition, Prentice Hall, Englewood Cliffs, NJ, 1988.

III.ME2 Current Issues and Research in Secondary School Mathematics

See the descriptions of **I.ME2** and **II.ME2**. Especially at Level III, participants need to consider their own teaching in terms of contemporary research and a problem solving approach. Each participant is assumed to be seeking ways to improve teaching by

Identifying instructional and curricular problems.

Hypothesizing and testing solutions for these problems.

Systematic evaluation.

Topics appropriate to this level include:

The use of calculators and computers in the classroom.

Graphing and analytic geometry.

Assessment of student achievement.

The role of parents.

An examination of recent findings published by the National Assessment of Educational Progress, the American Educational Research Association, and the NCTM.

Each issue is examined relative to:

Position statements of professional organizations.

Available research.

Observation of practices in schools.

Changes in algebra curriculum content.

The changing nature of geometry as a secondary school course.

Each participant writes a paper summarizing research findings and needed research in some area of mathematics education.

References (see also **I.ME2** and **II.ME2**)

Donald Dessart and Marilyn N. Suydam, *Classroom Ideas from Research on Secondary School Mathematics*, NCTM, Reston, VA, 1983.

The Mathematics Teacher, Special Issue on Gifted Students, NCTM, Reston, VA, April, 1983.

III.ME3 Diagnosis and Remediation in Secondary School Mathematics

See the description of **I.ME3** and **II.ME3**. At Level III, developmental models such as van Hiele (geometry), Hart/Kirkemann/Booth (variable, algebra, graphing), and the classical Piaget (reasoning) are used to provide a theoretical basis for analyzing learner's behavior. Participants interview individual learners on mathematics topics, analyze the learner's behavior, and prepare remediation programs for these individuals.

Language related difficulties.

Estimation.

Ratio and proportion.

The concept of variable.

Equation solving.

Graphs and their interpretation.

Geometrical visualization.

Theorem proving.

Problem solving.

References (see also **I.ME3** and **II.ME3**)

Lesley R. Booth, *Algebra: Children's Strategies and Errors*, Center for Science and Mathematics Education, NFER-Nelson, Windsor SL4 1DF, UK, 1984 (contains 81 references).

W.F. Burger and J.M. Shaughnessy, Characterizing the van Hiele levels of development in geometry, *Journal of Research in Mathematics Education*, 17 (1986) 31-48.

H. Ginsberg, *Children's Arithmetic: The Learning Process*, Van Nostrand, New York, 1977.

Kathleen Hart (ed.), *Children's Understanding of Mathematics: 11-16*, Murray, London. 1981.

Kathleen Hart, *Ratio: Children's Strategies and Errors*, Center for Science and Mathematics Education, NFER-Nelson, Windsor SL4 1DF, UK, 1984.

Alan Hoffer, Geometry is more than proof, *The Mathematics Teacher*, 74 (1981) 11-18.

Larry Sowder, *Didactics and Mathematics*, Oregon Mathematics Resource Project, Creative Publications, Palo Alto, CA, 1978.

III.ME4 The Secondary School Mathematics Curriculum

Transitions between different levels of schooling.

Incorporation of computers and calculators into the curriculum (including the use of symbol manipulators, e.g. muMath).

Under-represented topics (estimation, probability, statistics, mathematical modelling, discrete mathematics applications).

Testing as a determinant of curricular content (SAT, Advanced Placement, ACT, competency testing programs).

New curricular materials designed to meet current issues and problems.

Evaluation of curricular modifications.

References

Marilyn N. Suydam, *et al*, (eds.), *Alternative Courses for Secondary School Mathematics*, NCTM, Reston, VA, 1985.

Barry Barstow, Jill Hughes, Barry Kissane and Roly Mortlock, *40 Mathematical Investigations*, Mathematical Association of Western Australia, Perth, 1984.

E.G. Begle (ed.), *69th Yearbook of the National Society for the Study of Education: Mathematics Education*, University of Chicago Press, Chicago, 1970.

D.C. Carter (ed.), *Topics in Mathematics: Some Ideas for the Secondary Classroom*, Bell and Hyman, London, 1979.

M. Driscoll, *Research within Reach: Secondary School Mathematics*, NCTM, Reston, VA 1983.

H.B. Griffiths and A.G. Howson, *Mathematics: Society and Curricula*, Cambridge University Press, London, 1974.

Louise Grinstein and Brenda Michaels (eds.), *Calculus: Readings from The Mathematics Teacher*, NCTM, Reston, VA, 1977.

Bob McCreddin, Donna Carlton, Tony Edwards and Janet Hunt, *50 Mathematical Projects*, Mathematical Association of Western Australia, Perth, 1984.

Alan H. Schoenfeld, *Problem Solving in the Mathematics Curriculum*, MAA Notes Number 1, MAA, Washington DC, 1983.

R.J. Shumway, *Research in Mathematics Education*, NCTM, Reston, VA, 1980.

Christian R. Hirsch and Marilyn J. Zweng (eds.), *The Secondary School Mathematics Curriculum*, 1985 Yearbook, NCTM, Reston, VA.

III.ME5 Computers and Technology in Secondary School Mathematics[1]

See the descriptions for **I.M2** and **II.M2** . Many of the topics listed there are relevant to this level. Additional topics include:

Calculators--scientific and graphing.

Computer graphics. Spread sheets. Data bases.

Symbol processing software.

Identification of exemplary software by level, general mathematics through calculus.

Exemplary software by topic. Problem-solving software.

Innovative software like MACSYMA, TRUE BASIC, and MAPLE.

Using simulations and graphing software in the teaching of mathematics.

Experimental mathematics. Simulation.

Technology other than computers. Audio-visual aids.

Classroom management

References (see also **I.M2** and **II.M2**)

Harold Abelson, *Turtle Geometry*, MIT Press, Cambridge MA, 1981.

D. E. Arganbright,*Mathematics Applications of Electronic Spreadsheets*, McGraw-Hill, New York, 1984.

Barbara J. Bestgen and Robert E. Reys, *Films in the Mathematics Classroom*, NCTM, Reston, VA, 1982.

Gary G. Bitter, *Microcomputer Applications for Calculus*, Prindle, Weber and Schmidt, Boston, 1981.

James T. Fey (ed.), *Computing and Mathematics: The Impact on Secondary School Curricula*, NCTM, Reston, VA, 1984.

---

[1] See the statement on teaching computer science that follows the description of this course.

William P. Heck, Jerry Johnson, Robert J. Kansky and Dick Dennis, *Guidelines for Evaluating Computerized Instructional Materials*, 2nd edition, NCTM, Reston, VA, 1984.

A.G. Howson and J.P. Kahane, *The Influence of Computers on Mathematics and its Teaching*, Cambridge University Press (ICMI Study Series), London, 1986.

Robert James and Ray Kurtz, (eds.), *Science and Mathematics Education for the Year 2000 and Beyond*, School Science and Mathematics Association, Bowling Green, OH, 1985.

Peter Kelman, *et al*, *Computers in Teaching Mathematics*, Addison-Wesley, Reading, MA, 1983.

E.B. Koffman, D. Stemple and C.E. Wardle, Recommended curriculum for CS2, 1984, *Communications of the Association for Computing Machinery*, 28 (1985) 815-818.

Donald Small, John Hosack, and Kenneth Lane, Computer Algebra Systems in Undergraduate Mathematics, *The College Mathematics Journal*, 17 (1986) 423-441.

Herbert S. Wilf, Disk with the college education, *American Mathematical Monthly*, 89 (1982) 4-8.

Courseware

Cactus Plot, Cactus Plot Co., Tempe, AZ, 85281.

Computer Graphing Experiments, Addison Wesley, Menlo Park, CA, 94025.

Electronic Blackboard COMPress, Wentworth, NH, 03282.

Function Grapher by Ben Waits and Frank Demana, Ohio State University, Columbus, OH 1987.

Geometry, Borland International, Scotts Valley, CA, 95066

Geometry Supposers, Sunburst Communications, Pleasantville, NY, 10570.

Green Globs, Sunburst Communications, Pleasantville, NY, 10570.

Mathematics-Advanced, MECC, St. Paul, MN, 55126.

Microsoft muMath , Microsoft Corp., Redmond, WA, 98073.

Superplot, Edusoft, Berkeley, CA, 94702.

True Basic Mathematics Series, True Basic Inc., Hanover, NH, 03755.

TK Solver, Universal Technology Systems, Rockford, IL, 61101.

ADDITIONAL COMPETENCE IS NEEDED FOR TEACHING COMPUTER SCIENCE

Teachers who wish to prepare to teach secondary school courses in computer science need to develop extra competence. This may be acquired from sources other than formal course work. For recommendations for such preparation, see the publications of the Association for Computing Machinery.

Reference

Jean B. Rogers, David G. Moursund and Gerald L. Engel, Preparing precollege teachers for the computer age, *Communications of the Association for Computing Machinery*, 27 (1987) 195-200.

PROGRAMS FOR SUPERVISORS AND COORDINATORS

Mathematics teachers are extremely busy. They teach five or six classes a day, have about three preparations for different classes, work with up to 180 students each day, and often have responsibilities other than teaching mathematics. This leaves little time or energy for the things that make the difference between an adequate performance and an excellent one. There is not enough time for planning, evaluation, and consequent improvement of the program.

Teachers need the support of well-informed supervisors

The supervisor

- Provides an avenue for communication among teachers, administrators, and other appropriate groups.

- Defends the position of the mathematics program in the school.

- Identifies inservice needs, makes liaison with colleges and universities, and arranges programs.

- Helps to select textbooks.

- Ensures continuity of curriculum and instruction across the three levels and with institutions of higher learning.

- Makes recommendations of staffing.

- Lobbies more effectively and objectively than the classroom teachers.

We must adopt certification standards for supervisors commensurate with their wide responsibilities. These standards also provide for promotion without the complete loss to the system, so often suffered in the past, of classroom expertise.

The mathematics supervisor certificate can provide opportunities for professional teachers and can enable them to exert a subtle but direct pressure on school systems to provide the stewardship needed by school mathematics programs. Talented and appropriately educated supervisors of mathematics can provide the leadership needed by mathematics teachers at every level within a school system.

The lack of standards for mathematics supervisors in many states has led to some school systems hiring supervisory personnel for mathematics more for their political acumen than for what they know about mathematics and mathematics education. Certainly a supervisor of mathematics must have political skills as well as technical competence for the position. However, given the responsibilities for curriculum design and evaluation, a primary dependence on political answers is not sensible. Too much is at stake in the effective operation of a school mathematics program to settle merely for political skills.

A supervisor's background must span all levels of curricular and instructional responsibility. To be effective at the elementary school level a supervisor with mainly high school experience must understand the issues for instruction at the elementary level, must have an appreciation of the characteristics of young children as learners, and must be able to recognize effective teaching at that level. We assume that the supervisor will already have taken the masters program for Level II or Level III, and will have had at least three years teaching experience. We recommend courses **ME1** and **ME2** at Level I, II, or III, along with those of the potential supervisors' master's qualification.

A supervisor needs a broader perspective than the typical mathematics teacher (although, ideally, a mathematics teacher also would have such breadth and outlook). This general background ensures that the mathematics supervisor is speaking the same language as supervisors in other fields, understands the motivation of school administrators, and can communicate effectively with all of them.

We recommend the following courses, or their equivalents:

• Curriculum and Instruction for Special Populations, (including multicultural emphases).

• Child Psychology or Adolescent Psychology.

• General Curriculum for School Administrators.

• Introductory Supervision for School Administrators.

The supervisor certificate should reflect additional course work and experience beyond the master's program. The need for a course on 20th Century Topics in Mathematics depends on the recency and comprehensiveness of the candidate's mathematical preparation. The planned field experience is placed late in the program since it constitutes an internship allowing the prospective supervisor to use the skill and understanding acquired throughout the program and to evaluate his or her leadership ability.

DESCRIPTIONS OF SAMPLE COURSES FOR SUPERVISORS

S.1 Organization and Supervision of Mathematics Programs K-12

The course concerns the functions a supervisor must perform in order to promote an effective mathematics program. The course helps the supervisor prepare to:

Provide support services to teachers and courses.

Develop leadership in teachers.

Supervise beginning teachers.

Manage staff.

Design and conduct inservice education.

Work with school administrators, school boards, and peers in other curricular areas.

Plan and implement budgets.

Make management decisions and establish priorities among the supervisor's responsibilities.

References

Alan Osborne (ed.), *An In-service Handbook for Mathematics Education*, NCTM, Reston, Va, 1977.

J. Price and J.D. Gawronski (eds), *Changing School Mathematics: A Responsive Process*, NCTM, Reston, VA, 1981.

S.2 Curriculum and Instruction in Mathematics K-12

The intent is to bring issues and problems into perspective for the entire mathematics curriculum. The orientation is in terms of a single school system and its mathematics program five to ten years into the future. Where is it now, and what needs to be done? Among other topics of high priority, the course examines trouble points in curriculum and instruction. Currently these include:

Transition curriculum problems: elementary to middle school, middle school to high school, high school to college.

Serving the needs of both college bound and non-college bound students.

Incorporating the use of technology into the teaching of mathematics.

Motivating and educating teachers who do not like, or do not know, mathematics.

Working with parents and other lay people, in the interests of the school mathematics program.

Establishing appropriate liaison with people in industry, in business, in government, and in higher education.

The course must also attend to local and state requirements for mathematics programs. These include:

Textbook selection.

Developing and documenting the curriculum (graded course of study or curriculum guides).

Factors in planning for Public Law 42-192.

References

The professionals recruited into a program for supervisors of mathematics must develop a thorough, working knowledge of curriculum and instruction in mathematics education and be familiar with sources for ideas and approaches that have been available in the immediate past. Thus, the instructor for such a course has a responsibility for finding instructional materials that extends beyond the usual working with the library to provide the most recent published materials. (If it has already been published and is available commercially, it is probably too old to provide insights about the state of the art in curriculum and instruction.)

The instructor must make available the latest working drafts of recommendations for curriculum and instruction for the study and reactions of the participants in a supervisors program. As of January 1988, this means promoting the study and discussion of pace-setting documents such as:

Commission on Standards for School Mathematics, *Curriculum and Evaluation Standards for School Mathematics* , NCTM, Reston, VA, working draft 1987.

Project 2061: Phase I, AAAS, Washington, DC, integrated draft report.

Additionally, the instructor must strive to provide examples of current projects in curriculum and instruction that are under development. A good source for such project information is the annual directory of awards published by the National Science Foundation which can yield people to contact about current development efforts.

S.3 Evaluation of School Mathematics Programs

The supervisor of school mathematics must systematically collect and use information about the effectiveness of programs and the effectiveness of teachers. The following areas are reviewed, evaluated and criticized:

Student achievement (standardized tests, school-created tests).

Teacher performance.   Program evaluation.

Needs assessment.   Placement tests.

Testing of minimal competence.

The criteria considered are:

Suitability.   Effectiveness.

Reliability.   Cost.

Bias (sexual, racial, cultural).

Methods of designing comprehensive, cost-effective evaluation programs are considered.

The course also considers appropriate uses of assessment information, including who needs to know what, and how the information can best be used in making decisions about programs and individuals.

The supervisor needs a knowledge of statistics, including the usual measures of test reliability and validity, the fundamentals of evaluation design and statistical analysis. If participants do not have this background, it must be provided in the course.

References

Nina L. Ronshausen, *Facilitating Evaluation: The Role of the Mathematics Supervisor*, National Council of Supervisors of Mathematics, 1986.

Richard Wilkes, Dwight Coblentz and Dorothy Strong, Guidelines for Designing the In-Service Program, of *An In-Service Handbook for Mathematics Education*, A. Osborne (ed.), Reston, VA, NCTM, 1977.

David W. Wells and Mary Montgomery Lindquist, Evaluating Programs of Mathematics In-Service Education, of *An In-Service Handbook for Mathematics Education*, A. Osborne (ed.), Reston, VA, NCTM, 1977.

S.4 Planned Field Experience in Supervision of Mathematics

For a supervisor, this course is analogous to student teaching. Some experiences at levels different from the participant's previous background are included. The participant is apprenticed to an experienced supervisor of mathematics. Activities are selected from the full range of a supervisor's responsibilities, including:

Curriculum planning.

Implementing special and/or new programs.

Program, course, and teacher evaluation.

Support services for programs and for teachers.

Budget planning.

Counselling new teachers.

Developing teacher inservice programs.

Working with other supervisors of mathematics, department heads, principals, superintendents, parents, and school boards.

The planned field experience should be designed to include a wide variety of experiences reflecting the many different responsibilities that a supervisor might encounter. Of particular note, however, is field experience activity that focuses on the act of supervising an individual teacher almost in the sense of supervising a student teacher. The participant prepares a report that includes a reflective analysis of the experience acquired in the course.

References

D.C Berliner, Laboratory settings and the study of teacher education, *Journal of Teacher Education*, 36 (1985) 2-8.

J. Brophy and T.L. Good, Teacher behavior and student achievement, in *Handbook of Research on Teaching* (3rd edition), M.C. Wittrock (ed.), New York, 1984.

N. Gehrke, Rationale for field experiences in the professions, in *Exploratory Field Experiences in Teacher Education*, C. Webb et al (eds.), Association of Teacher Educators, Reston, Va, 1981.

T.L. Good, D.A. Grouws, and H. Ebmeier, *Active Mathematics Teaching*, Longman, White Plains, NY, 1983.

M. Haberman, Can common sense effectively guide the behavior of beginning teachers?, *Journal of Teacher Education*, 36 (1985) 32-35.

Marvin Minsky, *The Society of the Mind*, Simon & Schuster, New York, 1986.

Thomas R. Post (ed.), *Teaching Mathematics in Grades K-8: Research Based Methods*, Allyn and Bacon, Boston, 1988.

Max A. Sobel and Evan M. Maletsky, *Teaching Mathematics: A Sourcebook of Aids, Activities and Strategies*, (2nd edition), Prentice Hall, Englewood Cliffs, NJ, 1988.

D. Watts, Student teaching, in *Advances in Teacher Education: Vol. 3*, M. Haberman and J.M. Bachus (eds.) Ablex, Norwood, NJ, 1987.

S.5 Current Topics in Mathematics

The field of mathematics is undergoing change at a dramatic rate. New mathematics and different emphases are evolving and contributing to modifications of curricula at many colleges and universities. Discrete mathematics, combinatorics, mathematical modelling, and a variety of other relatively new topics are included in undergraduate programs at many institutions.

In order to appreciate the significance of curricular trends and acquire the perspective to influence change wisely, supervisors of school mathematics need to be familiar with major trends in mathematics. In some respects the emphasis is on a popularization of major content domains of recent significant mathematical developments. In other respects there is a significant encounter with problem-solving techniques and processes of mathematics.

It is not sufficient simply to talk or read about mathematics. The mathematical experiences are correlated with related readings concerning curricular trends and positions.

References

See **III.ME4**; e.g., Griffiths and Howson, and **III.M16**; e.g., the book edited by Lynn Arthur Steen.